SpringerBriefs in Water Science and Technology

SpringerBriefs in Water Science and Technology present concise summaries of cutting-edge research and practical applications. The series focuses on interdisciplinary research bridging between science, engineering applications and management aspects of water. Featuring compact volumes of 50 to 125 pages (approx. 20,000–70,000 words), the series covers a wide range of content from professional to academic such as:

- Timely reports of state-of-the art analytical techniques
- Literature reviews
- In-depth case studies
- Bridges between new research results
- Snapshots of hot and/or emerging topics

Topics covered are for example the movement, distribution and quality of freshwater; water resources; the quality and pollution of water and its influence on health; and the water industry including drinking water, wastewater, and desalination services and technologies.

Both solicited and unsolicited manuscripts are considered for publication in this series.

More information about this series at http://www.springer.com/series/11214

Samsul Ariffin Abdul Karim ·
Nur Fatonah Kamsani

Water Quality Index Prediction Using Multiple Linear Fuzzy Regression Model

Case Study in Perak River, Malaysia

 Springer

Samsul Ariffin Abdul Karim
Fundamental and Applied Sciences
Department and Centre for Smart Grid
Energy Research (CSMER)
Institute of Autonomous System
Universiti Teknologi PETRONAS
Seri Iskandar, Perak, Malaysia

Nur Fatonah Kamsani
Fundamental and Applied Sciences
Department
Universiti Teknologi PETRONAS
Seri Iskandar, Perak, Malaysia

ISSN 2194-7244   ISSN 2194-7252 (electronic)
SpringerBriefs in Water Science and Technology
ISBN 978-981-15-3484-3   ISBN 978-981-15-3485-0 (eBook)
https://doi.org/10.1007/978-981-15-3485-0

This Springer imprint is published by the registered company Springer Nature Singapore Pte Ltd.
The registered company address is: 152 Beach Road, #21-01/04 Gateway East, Singapore 189721, Singapore

Preface

This study discusses a new method to calculate the Water Quality Index (WQI). The multiple linear regression based fuzzy model is proposed. The existing and standard WQI formulation is proposed by the Department of Environmental (DOE), Malaysia. We modify the formula by embedding the fuzzy multiple regression by considering the spreading for each data. There are about six parameters that need to be collected such as concentration of hydrogen ion (pH), Biological Oxygen Demand (BOD in mg/l), Chemical Oxygen Demand (COD in mg/l), Ammoniacal Nitrogen (AN in mg/l), Suspended Solid (SS in mg/l) and Dissolved Oxygen (DO in mg/l). We use the data collected at Perak River, State of Perak, Malaysia from year 2013 until year 2017.

We would like to thank Universiti Teknologi PETRONAS (UTP), Malaysia and Universitas Islam Riau (UIR), Pekanbaru, Riau, Indonesia for providing the financial support through International Collaborative Research Funding (ICRF) Grant No: 015ME0-037. Thank you very much to Department Chair, Assoc. Prof. Dr. Hanita Daud, our team members Assoc. Prof. Dr. Mahmod Othman from UTP and Dr. Evizal Abdul Kadir from UIR for their feedback and encouragement that have made this project is completed within the given time frame.

Seri Iskandar, Malaysia

Samsul Ariffin Abdul Karim
Nur Fatonah Kamsani

Acknowledgements

This study is fully supported by Universiti Teknologi PETRONAS (UTP) and Universitas Islam Riau (UIR), Pekanbaru, Indonesia for the financial support received in the form of a research grant: International Collaborative Research Funding (ICRF): 015ME0-037 for the period 1 August 2018 until 31 January 2020. We wish to thank both institutions that have made this book project possible for its completion. Much of this work were done when the second author is working as Research Assistant (RA) under the research grant from 16 September 2019 until 16 January 2020. The data used in this study was supplied by Department of Environmental (DOE), Malaysia, Water and Marine Division, River Water Quality Section. Therefore, we would like to thank DOE, Malaysia.

Contents

About the Authors

Samsul Ariffin Abdul Karim is a senior lecturer at Fundamental and Applied Sciences Department, Universiti Teknologi PETRONAS (UTP), Malaysia. He has been in the department for more than eleven years. He obtained his B.App.Sc., M.Sc. and Ph.D. in Computational Mathematics & Computer Aided Geometric Design (CAGD) from Universiti Sains Malaysia (USM). He had 20 years of experience using Mathematica and MATLAB software for teaching and research activities. His research interests include curves and surfaces designing, geometric modeling and wavelets applications in image compression and statistics. He has published more than 120 papers in Journal and Conferences as well as seven books including two research monographs and three Edited Conference Volume and 25 book chapters. He was the recipient of Effective Education Delivery Award and Publication Award (Journal & Conference Paper), UTP Quality Day 2010, 2011 and 2012 respectively. He is Certified WOLFRAM Technology Associate, Mathematica Student Level. He has published three books with Springer including **Sustaining Electrical Power Resources through Energy Optimization and Future, Engineering, Springer Briefs in Energy, 2018**.

 **Nur Fatonah Kamsani** graduated in Bachelor of Science (Hons) Computational Mathematics from Universiti Teknologi MARA (UiTM), Malaysia in 2019. Currently she is working as Research Assistant in Universiti Teknologi PETRONAS (UTP). She is interested in programming and her research interests are modeling and data analysis. She has experienced in website development.

Chapter 1
Introduction

Abstract This chapter discuss the introduction on the subject matter including some related literature reviews as well as the motivation of the study. The background of the study is presented in details.

Keywords Water quality · Fuzzy · Crisp · Fuzzy regression · River

1.1 Introduction

Water is crucial in our life as it is the main source of our ecosystem. All living organisms need water to grow and survive. As we know, our mother earth is full of water. Two-thirds is water and the other one-third is land. Drinking water or freshwater, however, is still scarce as it consists of only 0.3% of that body. Yet we do not know how to appreciate it as the occupant of Earth. As being said by Albert Szent Gyorgyi: "Water is a life's mater and matrix, mother and medium. There is no life without water". Imagine one day when we have no more clean water supply and need to drink from the water from the Fig. 1.1. Definitely, we will be suffering since

Fig. 1.1 Image of polluted water [33]

there are no clean water even for drink and to take bath. This is unthinkable event for the next fifty years.

In order for us to be able to use the water safely without any health issues later, the water quality need to be determined first before it can be used. To do this, we need to check the quality index of the water. This is where water quality index (WQI) is used and it has been proposed by the Department of Environmental (DOE) around the globe. For instance, in Malaysia, Malaysia Department of Environmental is responsible to provide the WQI for all rivers in Malaysia as well as for drinking and mineral waters etc. This to ensure that the WQI is updated for each river and the residents must be provide the details if some rivers are polluted by chemical or others.

Water quality is a phrase to describe the chemical, physical and biological characteristics of water. There are initially nine parameters included in calculating WQI which are dissolved oxygen, faecal coliforms, pH, biochemical oxygen demand (BOD), nitrates, phosphates, temperature, turbidity, and total solids [5]. Based on the studies of [5] and [38], as well as the standard prepared by DOE, six parameters as mentioned below are enough to represent the WQI model. Defining 'good' or 'bad' water quality is not as simple as it seems because it depends on the context in which it is used. A simple example would be that water good enough to wash your car may not be good enough for drinking. However, in general there are some standard ways for measuring water quality.

Regression analysis was first developed in the latter part of the 19th century by Sir Francis Galton. Regression analysis has become one of the standard data analysis tools. Its popularity comes from various fields. From its analysis, the mathematical equation can explain the relationship between dependent and independent variables. It has been commonly used in applied sciences, economics, engineering, computer sciences, social sciences, and other areas. Multiple regression is a flexible data analysis approach that can be useful when analyzing a quantitative variable in relation to any other variables.

Fuzzy multiple linear regression is a method which is used to handle the uncertainty in variables used in the multi-linear regression method. Conventional regression cannot accommodate non-crisp or linguistic visual inspection results. Other than that, fuzzy regression provides an effective way to cope with such fuzzy data or linguistic variables. This is the main advantages of fuzzy method in regression or forecasting i.e. it can cater the uncertainties of the data.

Thus, in this study, we will be investigating the effect of all six parameters toward the developing of new fuzzy multivariate regression analysis to determine the relationship between parameters. The parameters are pH, chemical oxygen demand (COD), ammoniacal nitrogen, dissolved oxygen, suspended solid and biochemical oxygen demand (BOD).

The main differences between our proposed method and [4] is that, in [4]'s work, the author used only several parameters such as water temperature, TDS, BOD and DO to define the water quality. For their first case study, he monitored TDS and temperature parameters in different water depth, and then applied the method to the data and with secondary data from different organizations that used to develop a suitable model with which it is possible to predict the variation of TDS in the future. While in our study, we used collected data from DOE Malaysia and used six parameters as mentioned before in order to determine the water quality. Then, we applied fuzzy linear regression to assess the water pollution levels. We believe that our proposed model is more suitable in predicting the water quality index and the class of the river.

1.2 Water Supplies

Water is an essential element in the conservation of life and is required for all creatures to survive. Water covers 71% of the surface of the earth and the presence of water makes it possible for different life forms to breed. However, about 98% of world's surface water exists as salt water in the ocean and the seas. Only 2% of the world's surface water is fresh water found in lakes, inland seas and rivers. Unfortunately, 99% of these fresh waters are hard to access as they are either frozen in glaciers or ice caps, kept as soil moisture, or stored too deep in water tables to access. So, this means that there is only 1% of world's total fresh water supply is available for consumption by humans, animals and irrigation.

Usually in Malaysia, we expect to have access to clean water every time we turn on the tap. For most of our lives, we have been thinking in these ways and we are expecting the same in the future. But, how many of us do actually know where our water comes from and where does it go? A better understanding of types of water will help us better in managing our sources of water.

There are various types of water and water supplies such as surface water and groundwater, wastewater and stormwater. Surface waters include streams, rivers, lakes, reservoirs, and wetlands, while groundwater is the water beneath the earth. As we know wastewater is used water mostly developed from combination of human activities such as agricultural activities, domestic, industrial and also from sewer inflow. From U.S. EPA, stormwater is rainwater or melted snow that runs off streets, lawns and other sites. When the soil is unable to absorb the stormwater, it will result in surface runoff draining into the rivers, lakes and oceans.

Malaysia has been blessed with abundant rainfall averaging 3,000 mm of rainfall per year, which leads to an estimated annual water resource of around 900 billion m^3. Our river systems consist of more than 100 river systems in Peninsular Malaysia and over 50 river systems in Sabah and Sarawak. The main source of water supply is from surface water which is rivers (originate and flow from highlands), as Malaysia obtains nearly 97% from surface water sources and another 3% from underground sources. However, not all surface water available is safe to use. Over the past three decades, rapid population, public and industrial activities, and rural-urban migration have resulted in heavy water demand. These activities will also result in draining more wastewater into the local streams and rivers. Thus, it will affect and pollute our main sources of water which is surface water.

1.3 Literature Review

With two-thirds of the earth's surface covered by water and 75% of the human body, it is obvious that water is one of the primary elements responsible for life on earth. Water flows through the earth just as it flows through the human body, transporting, dissolving, replenishing nutrients and organic matter while transporting waste material. As been mentioned before in the earlier section, river is one of the main supplies of our freshwater. It carries water and nutrients to areas all around the earth. However, day by day our river is at risk of pollution. Activities based on water and land, such as maritime industrial activities, conventional types of agriculture and mining, not only affect the quality of local rivers but lakes, streams, and groundwater as well. In Malaysia, major pollution is coming from domestic waste, industrial effluents, land clearance with suspended solid (SS) as the major source contributing up to 42% to poorly planned land development, 30% from biological oxygen demand (BOD) due to industrial waste and 28% from ammoniacal nitrogen (NH_3-NL) attributed from domestic sewage disposal and animal farming activities.

There are various approaches taken in order to determine the water quality of river. According to Cude [13], revisions of water quality indices have been a continuing concern, as the various studies show; these studies have shown new approaches while providing new tools to establish other indices [14, 17, 22]. One of the earliest studies by Horton [20], he proposed the first formal water quality index (WQI) which is to study general indices, selecting and weighting parameters. He selected 10 most frequently used water quality variables such as dissolved oxygen (DO), pH, coliforms, specific conductance, alkalinity, and chloride, etc. in his research, and has been widely used and accepted in European, African and Asian countries. Later, the group of Brown [8] developed a new WQI similar to [20]'s index in 1970, which was based on weights to individual parameters. Among the first influential comparisons of water quality indices were [23], followed by [27], who updated water quality indices used in the United States, in addition to publishing a detailed discussion on environmental indices practices and theories. From Canter [9], he stated that initially WQI was designed to include nine parameters including dissolved oxygen, faecal coliforms, pH, biochemical oxygen demand (BOD), nitrates, phosphates, temperature, turbidity, and total solids. However, based on [5] and [38] and also from the standard prepared by DOE, only six out of nine parameters mentioned above are enough to represents the WQI model.

Regression analysis is a powerful statistical method for examining the relationship between two or more variables of interest. Although there are many types of regression analysis, the root of the model is still on the influence of one or more independent variables on one dependent variables. One of the regression analysis models is multiple regression analysis. According to [6], multiple linear regression (MLR) modeling is a powerful technique that is commonly used in agricultural research. There are some studies that used multiple linear regression in environmental science i.e. [2] used MLR to choose the best model in forecasting WQI in Manjung river, located in Perak River Basin. Meanwhile, Bhavyashree et al. [6] used MLR and fuzzy regression to predict mulberry leaf yield, and [31] used MLR to evaluate the influence of non-biological pollutant. One downside of the MLR is that the underlying relationship is expected to be precise and crisp, as it offers a precise response value for a series of explanatory variables values. The underlying relationship, however, is not a crisp function of a given form in a realistic situation; it contains some vagueness or inaccuracy. So, fuzzy set theory is an effective approach to analyzing data with uncertainty and imprecision [36].

Fuzzy regression analysis has been widely used in various fields as it has the ability to cope with problems in which human experts rely on subjective judgment results which can be non-crisp or linguistic [28]. It was stemming from [37] thought that fuzzy that was able to deal with ambiguity. According to [30], the uncertainties related by using fuzzy membership with values ranging from 0 to 1 to form an applicable fuzzy set instead of 0–100 scale used in traditional WQI rating curves. Tanaka et al. [4] introduced the fuzzy regression analysis more than thirty years ago. They describe the fuzzy uncertainty of dependent variables in the linear regression model with the fuzziness of response functions or regression coefficients. A few years later, Sakawa and Yano [29] and D'Urso and Gastaldi [16] generalized the idea of a fuzzy linear regression. Basically, it is possible to classify the fuzzy regression models into two classes. The first class is based on the concept of possibility [4, 11] whereas the second class is the least square method [15, 16, 34, 35]. D'Urso and Gastaldi [15] introduced a new approach to an analysis of fuzzy linear regression where they developed a linear adaptive model of fuzzy regression. It is based on two linear models like a model of center regression and a model of spread regression. Krätschmer [21] developed new fuzzy linear regression by incorporated the physical vagueness of the items involved in the form of fuzzy variables data where the type of single ordinary equation has been generalized in linear regression models. Other than that, he suggested that ordinary least-squares method has greater flexibility for modelling and estimation.

There is a tendency in fuzzy regression that the higher the values of independent variables, the wider the width of the estimated dependent variables [31]. This causes the accuracy of the fuzzy regression model built by the least square method to decrease. From the work of Choi and Buckley [12], they proposed to construct the fuzzy regression model using least square absolute deviation estimators and investigate the performance of the fuzzy regression models with respect to a certain error measure. Through their simulation studies and examples, it is proven that their model produced less error than other existing researches of fuzzy regression model by other authors that used least square method when the data contain outliers.

Due to the tedious fuzzy arithmetic, many of the existing fuzzy regression models require substantial computations. In 1982, Asai et al. [4] proposed a quite popular and useful regression model. However, this model is limited only to symmetric triangular fuzzy numbers. Chang and Lee [10] overcome this problem by developing a fuzzy least-squares regression model. Later, Pan et al. [28] came up with an idea of introducing matrix-driven multivariate fuzzy linear regression in order to improve the computational efficiency faced by Chang and Lee [10]. The approach was successfully tested to estimate bridge performance in the engineering study. The matrix-driven fuzzy regression uses matrix-algebra to present a multiple fuzzy linear regression. A combination of fuzzy data and crisp data can be handled by the model. Compared to other similar fuzzy regression models, this model is also intuitive and easy to implement. Abdullah and Zakaria [1] extended Pan et al. [28]'s research by simplifying the matrix-driven fuzzy linear regression into nine step computation procedures. Other than estimating bridge performance, this method also has been successfully tested to car sales volume, dimension of health-related quality of life and also taxation analysis [1, 19, 31] (Table 1.1).

Table 1.1 Literature review

Reference	Relevant findings
Cude [13]	Revisions of water quality indices have been a continuing concern
Horton [20]	Proposed the first formal water quality index (WQI) which is to study general indices, selecting and weighting parameters
Brown [8]	Developed a new WQI similar to [20]'s index in 1970, which was based on weights to individual parameters.
Landwehr and Deininger [23], Ott [27]	Updated water quality indices used in the United States, in addition to publishing a detailed discussion on environmental indices practices and theories
Canter [9]	Stated that initially WQI was designed to include nine parameters including dissolved oxygen, faecal coliforms, pH, biochemical oxygen demand (BOD), nitrates, phosphates, temperature, turbidity, and total solids
Bakar et al. [5], Zainordin et al. [38]	Only six out of nine parameters mentioned above are enough to represents the WQI model
Bhavyashree et al. [6]	• Multiple linear regression (MLR) modeling is a powerful technique that is commonly used in agricultural research • Used MLR and fuzzy regression to predict mulberry leaf yield
Ahmad and Yahaya [2]	Used MLR to choose the best model in forecasting WQI in Manjung river, located in Perak River Basin
Sousa et al. [31]	Used MLR to evaluate the influence of non-biological pollutant
Yen and Langari [36]	Fuzzy set theory is an effective approach to analyzing data with uncertainty and imprecision
Pan et al. [28]	Introduced a matrix-driven multivariate fuzzy linear regression as part of their efforts to improve computational efficiency

(continued)

Table 1.1 (continued)

Reference	Relevant findings
Zadeh [37]	It was stemming from Zadeh thought that fuzzy that was able to deal with ambiguity
Sii et al. [30]	The uncertainties related by using fuzzy membership with values ranging from 0 to 1 to form an applicable fuzzy set instead of 0–100 scale used in traditional WQI rating curves
Tanaka et al. [4]	• Introduced the fuzzy regression analysis more than thirty years ago • They describe the fuzzy uncertainty of dependent variables in the linear regression model with the fuzziness of response functions or regression coefficients
D'Urso et al. [15, 16]	• The first class is based on the possibility concept and the second class is the least-squares approach • Introduced a new approach to an analysis of fuzzy linear regression where they developed a linear adaptive model of fuzzy regression
Krätschmer [21]	Developed new fuzzy linear regression by incorporated the physical vagueness of the items involved in the form of fuzzy variables data
Choi and Buckley [12]	Proposed to construct the fuzzy regression model using least square absolute deviation estimators and investigate the performance of the fuzzy regression models with respect to a certain error measure
Chang et al. [10]	Developed a fuzzy least-squares regression model but for their studies, the regression coefficient are derived from a nonlinear programming problem that requires considerable computations.
Abdullah and Zakaria [1]	Extended [28]'s research by simplifying the matrix-driven fuzzy linear regression into nine step computation procedures

1.4 Objective of the Study

The research problem statements are summarized below:

(1) What are relationships between all six parameters with WQI?
(2) How to derive the Fuzzy multivariate regression analysis?
(3) What is the fitness measurement of the proposed fuzzy model?

The main objective of the studies are:

(1) To investigate the relationships between all six parameters with WQI.
(2) To derive the Fuzzy multivariate regression analysis.
(3) To validate the proposed fuzzy multivariate model and compare with existing method.

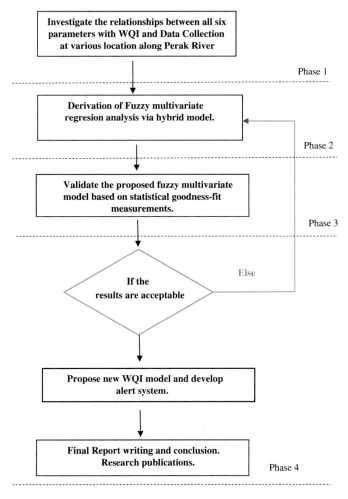

Fig. 1.2 Flowchart of the study

Figure 1.2 shows the flowchart of the study.

References

1. Abdullah L, Zakaria N (2012) Matrix driven multivariate fuzzy linear regression model in car sales. J Appl Sci (Faisalabad) 12(1):56–63
2. Ahmad F, Yahaya S (2017) First-order interaction multiple regressions model on water quality index in Manjung River and its tributaries
3. Asadollahfardi G (2015) Water quality management: assessment and interpretation. Springer, Berlin

4. Asai HTSUK, Tanaka S, Uegima K (1982) Linear regression analysis with fuzzy model. IEEE Trans. Systems Man Cybern 12:903–907
5. Bakar AAA, Pauzi AM, Mohamed AA, Sharifuddin SS, Idris FM (2018) Preliminary analysis on the water quality index (WQI) of irradiated basic filter elements. In: IOP conference series: materials science and engineering, vol 298, no 1. IOP Publishing, p 012005
6. Bhavyashree S, Mishra M, Girisha GC (2017) Fuzzy regression and multiple linear regression models for predicting mulberry leaf yield: a comparative study. Int J Agric Stat Sci 13(1):149–152
7. Boyd CE (2020) Water quality: an introduction, 2nd edn. Springer, Berlin
8. Brown RM, McClelland NI, Deininger RA, Tozer RG (1970) A water quality index—do we dare
9. Canter LW (2018) Environmental impact of water resource projects. CRC Press
10. Chang PT, Lee ES (1996) A generalized fuzzy weighted least-squares regression. Fuzzy Sets Syst 82(3):289–298
11. Chen YS (2001) Outliers detection and confidence interval modification in fuzzy regression. Fuzzy Sets Syst 119(2):259–272
12. Choi SH, Buckley JJ (2008) Fuzzy regression using least absolute deviation estimators. Soft Comput 12(3):257–263
13. Cude CG (2001) Oregon water quality index a tool for evaluating water quality management effectiveness 1. JAWRA J Am Water Resour Assoc 37(1):125–137
14. Dinius SH (1987) Design of an index of water quality 1. JAWRA J Am Water Resour Assoc 23(5):833–843
15. D'Urso P, Gastaldi T (2000) A least-squares approach to fuzzy linear regression analysis. Comput Stat Data Anal 34(4):427–440
16. D'Urso P, Gastaldi T (2001) Linear fuzzy regression analysis with asymmetric spreads. In: Advances in classification and data analysis. Springer, Berlin, Heidelberg, pp 257–264
17. Dojlido J, Raniszewski J, Woyciechowska J (1994) Water quality index applied to rivers in the Vistula river basin in Poland. Environ Monit Assess 33(1):33–42
18. Duca G (2014) Management of water quality in Moldova. Springer International Publishing, Cham
19. Hidayah Mohamed Isa N, Othman M, Karim SAA (2018) Multivariate matrix for fuzzy linear regression model to analyse the taxation in Malaysia. Int J Eng Technol 7(4.33):78–82. http://dx.doi.org/10.14419/ijet.v7i4.33.23490
20. Horton RK (1965) An index number system for rating water quality. J Water Pollut Control Fed 37(3):300–306
21. Krätschmer V (2006) Least-squares estimation in linear regression models with vague concepts. Fuzzy Sets Syst 157(19):2579–2592
22. Kung HT, Ying LG, Liu YC (1992) A complementary tool to water quality index: fuzzy clustering analysis 1. JAWRA J Am Water Resour Assoc 28(3):525–533
23. Landwehr, J. M., & Deininger, R. A. (1976). A comparison of several water quality indexes. Journal (Water Pollution Control Federation), 954–958
24. Li Y, Nzudie HLF, Zhao X, Wang H (2020) Addressing the uneven distribution of water quantity and quality endowment: physical and virtual water transfer within China. In: SpringerBriefs in water science and technology. Springer, Singapore
25. Marcello B, George T (2013) Water quality modelling for rivers and streams. Springer, Water Science and Technology Library
26. Mohammadpour R, Shaharuddin S, Chang CK, Zakaria NA, Ab Ghani A, Chan NW (2015) Prediction of water quality index in constructed wetlands using support vector machine. Environ Sci Pollut Res 22(8):6208–6219
27. Ott W (1978) Water quality indices: a survey of indices used in the United States, vol 1. Environmental Protection Agency, Office of Research and Development, Office of Monitoring and Technical Support
28. Pan NF, Lin TC, Pan NH (2009) Estimating bridge performance based on a matrix-driven fuzzy linear regression model. Autom Constr 18(5):578–586

29. Sakawa M, Yano H (1992) Multiobjective fuzzy linear regression analysis for fuzzy input-output data. Fuzzy Sets Syst 47(2):173–181
30. Sii HI, Sherrard JH, Wilson TE (1993) A water quality index based on fuzzy set theory. In: Environmental engineering-conference. American Society of Civil Engineers, pp 1727–1727
31. Sousa SIV, Martins FG, Pereira MC, Alvim-Ferraz MCM, Ribeiro H, Oliveira M, Abreu I (2010) Use of multiple linear regressions to evaluate the influence of O3 and PM10 on biological pollutants. Int J Environ Sci Eng 2(2):107–112
32. Tsuzuki Y (2014) Pollutant discharge and water quality in urbanisation. In: SpringerBriefs in water science and technology. Springer International Publishing, Cham
33. Water Pollution (2011) Retrieved from http://malaysianh2o.blogspot.com/2011/04/water-pollution.html
34. Wu HC (2003) Fuzzy estimates of regression parameters in linear regression models for imprecise input and output data. Comput Stat Data Anal 42(1–2):203–217
35. Xu R, Li C (2001) Multidimensional least-squares fitting with a fuzzy model. Fuzzy Sets Syst 119(2):215–223
36. Yen J, Langari R (1999) Fuzzy logic: intelligence, control, and information, vol 1. Prentice Hall, Upper Saddle River, NJ
37. Zadeh LA (1965) Fuzzy sets. Inf Control 8(3):338–353
38. Zainordin NS, Ramli NA, Elbayoumi M (2017) Distribution and temporal behaviour of O3 and NO2 near selected schools in Seberang Perai, Pulau Pinang and Parit Buntar, Perak, Malaysia. Sains Malays 46(2):197–207
39. Zali MA, Retnam A, Juahir H, Zain SM, Kasim MF, Abdullah B, Saadudin SB (2011) Sensitivity analysis for water quality index (WQI) prediction for Kinta River, Malaysia. World Appl Sci J 14:60–65

Chapter 2
Fuzzy Multiple Linear Regression

Abstract This chapter discusses the construction of the fuzzy multiple linear regression by embedding the spreading value and the crisp in the formulation. We begin by discussing the standard multiple linear regression. Then a method to calculate the spreading value is discussed in details. We propose a new method to calculate the spreading value. We establish the fuzzy multiple linear regression formulation by using fuzzy triangular number with symmetric property. This to ensure that, the calculation be made faster compared with asymmetric fuzzy triangular number.

Keywords Symmetric · Spreading value · Triangular fuzzy number · Least square · Fuzzy coefficient

2.1 Introduction

Regression analysis has been widely used in numerous fields of research and standard tools in analyzing data. Basically, it is used to explain the statistical relationship between explanatory (independent) and response (dependent) variables. The use of statistical linear regression is constrained by certain premises that error terms are mutually independent and distributed identically [16]. The statistical regression model also known as Multiple Linear Regression can only be implemented if the data given is distributed in accordance with the statistical model and the relationship between the explanatory and response variables in crisp. The underlying relationship, however, is not a crisp function of a given form in a realistic situation; it contains some vagueness or inaccuracy [4]. Therefore, some vital information can be lost by assuming a crisp relationship [15]. So, in order to handle the ambiguity, fuzzy linear regression model was introduced. Here, explanatory variables are assumed to be precise. On the other hand, fuzzy linear regression is suitable for coping with such fuzzy data or linguistic variables. It was stemming from [17] thought that fuzzy uncertainty as ambiguity and vagueness [1, 6, 12].

Tanaka et al. [13], was first introduced the fuzzy linear regression model by using linear programming problem to determine the regression coefficient as fuzzy numbers. Later, many researchers such as [14] and [7] has extended the idea of fuzzy linear regression. Because of the tedious fuzzy arithmetic, many of the current

© The Author(s), under exclusive licence to Springer Nature Singapore Pte Ltd. 2020   11
S. A. A. Karim and N. F. Kamsani, *Water Quality Index Prediction Using Multiple Linear Fuzzy Regression Model*, SpringerBriefs in Water Science and Technology, https://doi.org/10.1007/978-981-15-3485-0_2

fuzzy regression models require a lot of computations. Tanaka et al.'s [13] proposed regression model is indeed quite common and useful; however, this model is limited to symmetrical triangular fuzzy numbers. In order to overcome this constraint, [5] developed a fuzzy least-square regression model, but the regression coefficient in their model is derived from a nonlinear programming problem that requires considerable computations. Thus, as part of their efforts to improve computational efficiency, Pan et al. [11] proposed a matrix-driven multivariate fuzzy linear regression. The proposed approach can handle the functions of asymmetric and symmetric triangular membership. Other than that, this method has been successfully tested in engineering study of estimating bridge performance.

2.2 Multiple Linear Regression

Multiple Linear Regression model is a statistical technique that uses multiple explanatory variables to predict the outcome of a response variable. Multiple linear regression (MLR) is aimed at modeling the linear relationship between the explanatory (independent) variables and the response (dependent) variable. Multiple regression is basically the extension of ordinary least-squares (OLS) regression involving more than one explanatory variable. This model uses two approaches to generalize the simple linear regression. This allows more than one explanatory variable to rely on the mean function $E(y)$.

Let y denotes the dependent variable that is linearly related to k independent variables $X_1, X_2, \ldots, X_n$ through the parameters $\beta_1, \beta_2, \ldots, \beta_k$. The formula of MLR is as follows:

$$y = X_1\beta_1 + X_2\beta_2 + \cdots + X_k\beta_k + \varepsilon$$

The parameters $\beta_1, \beta_2, \ldots, \beta_k$ are the regression coefficients related with $X_1, X_2, \ldots, X_n$ respectively and ε is the element of random error which represents the difference between the observed and fitted linear relationship. The explanation of the formula is shown below, where the n equations can be written as

$$\begin{bmatrix} y_1 \\ y_2 \\ \vdots \\ y_n \end{bmatrix} = \begin{bmatrix} x_{11} & x_{12} & \ldots & x_{1k} \\ x_{21} & x_{22} & \cdots & x_{2k} \\ \vdots & \vdots & \vdots & \vdots \\ x_{n1} & x_{n2} & \cdots & x_{nk} \end{bmatrix} \begin{bmatrix} \beta_0 \\ \beta_1 \\ \vdots \\ \beta_k \end{bmatrix} + \begin{bmatrix} \varepsilon_1 \\ \varepsilon_2 \\ \vdots \\ \varepsilon_n \end{bmatrix}$$

or

$$y = X\beta + \varepsilon$$

where $y = [y_1, y_2, \ldots, y_n]'$ is a $n \times 1$ vector of n observation on dependent variable,

$$X = \begin{bmatrix} x_{11} & x_{12} & \ldots & x_{1k} \\ x_{21} & x_{22} & \cdots & x_{2k} \\ \vdots & \vdots & \vdots & \vdots \\ x_{n1} & x_{n2} & \cdots & x_{nk} \end{bmatrix}$$ is a $n \times k$ matrix of n observations on each of the k

explanatory variables, $\beta = [\beta_1, \beta_2, \ldots, \beta_n]'$ is $k \times 1$ vector of regression coefficients and $\varepsilon = [\varepsilon_1, \varepsilon_2, \ldots, \varepsilon_n]'$ is $n \times 1$ vector of random error components or disturbance term.

Take the first row of X as $[1, 1, \ldots, 1]'$ if there is an intercept term. So that

$$X = \begin{bmatrix} 1 & x_{11} & x_{12} & \ldots & x_{1k} \\ 1 & x_{21} & x_{22} & \ldots & x_{2k} \\ \vdots & \vdots & \vdots & \vdots & \vdots \\ 1 & x_{n1} & x_{n2} & \ldots & x_{nk} \end{bmatrix}$$. There are $[k-1]$ explanatory variables and one intercept

term in this case.

For MLR model, it used least squares methods to get the value of β. A general approach for estimating the vector coefficient of regression is to minimize

$$\sum_{i=1}^{n} M(\varepsilon_i) = \sum_{i=1}^{n} M(y_i - x_{i1}\beta_1 - x_{i2}\beta_2 - \cdots - x_{ik}\beta_k) \tag{2.1}$$

for properly selected function M.

Let B be the set of all β vectors possible. If there is no additional information, then B is the real Euclidean space k-dimensional. The goal is to find a vector $b' = [b_1, b_2, \ldots, b_k]$ from B that minimizes the sum of ε_i's squared deviations, i.e.,

$$S(\beta) = \sum_{i=1}^{n} \varepsilon_i^2 = \varepsilon'\varepsilon = (y - X\beta)'(Y - X\beta) \tag{2.2}$$

for given y and X. As $S(\beta)$ is a valued, convex and differentiable variable, a minimum will always exist.

Write (2.2) as $S(\beta) = y'y + \beta'X'X\beta - 2\beta'X'y$.

Differentiate (2.2) with respect to β yields

$$\frac{\partial S(\beta)}{\partial \beta} = 2X'X\beta - 2X'y$$

$$\frac{\partial^2 S(\beta)}{\partial \beta \partial \beta'} = 2X'X$$

The normal equation is

$$\frac{\partial S(\beta)}{\partial S} = 0 = X'X\beta = X'y \tag{2.3}$$

Multiply both sides of Eq. (2.3) by the inverse of $X'X$ to solve the normal equations.

Therefore, the least square estimator of β is given as

$$\beta = (X'X)^{-1}X'y \qquad (2.4)$$

2.3 Fuzzy Triangular Number

In situations of uncertainty, the fuzzy number is defined and applied to real-world science and engineering issues. Researchers around the globe have long developed different concepts that connect Fuzzy with most of the Mathematics field and introduced Fuzzy real line [8], Fuzzy topology [9], Fuzzy trigonometry [3] etc. Then, they have identified fuzzy numbers and found that there were more developments in application than fuzzy subsets. Fuzzy numbers were used to get better results when it comes to problems involving decision making and analysis. From Anand and Bharatraj's [2] studies, they stated that the behavior and properties of triangular fuzzy numbers resemble Pythagorean triples.

Definition 2.1 *Triangular Fuzzy Number* A fuzzy number $\tilde{A} = (a, b, c)$ is called triangular fuzzy number if its membership function is given by:

$$\mu_{\tilde{A}}(x) = \begin{cases} 0, x < a, x > c \\ \frac{x-a}{b-a}, x \in [a, b] \\ \frac{c-x}{c-b}, x \in [b, c] \end{cases}$$

In Anand and Bharatraj [2] works, they stated the concept of triangular fuzzy number as follows.

c, j and a are natural numbers and satisfy Pythagorean identity $c^2 + j^2 = a^2$. It also can be called Pythagorean triple or Pythagorean Triangle.

In terms of algebraic identities, without loss of generality, we can always find two natural numbers j_1 and j_2 such that the following are satisfied:

- $g.c.d(j_1, j_2) = 1$ where g.c.d is greatest common divisor.
- $c = j_2^2 - j_1^2$, $j = 2j_1 j_2$ and $a = j_2^2 + j_1^2$.

- $j_2 > j_1 > 0$.

Arithmetic operation between TFNs

Some important properties of operations on TFN are summarized below [18]:

i. The results from addition or subtraction between two TFNs are also TFNs
ii. However, the results from multiplication or division between two TFNs, although they are FNs, are not always TFNs.

Fig. 2.1 Symmetrical fuzzy regression coefficient

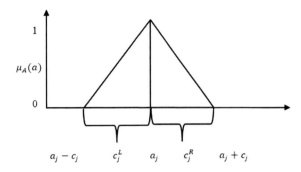

The general form of the model can be written as in (2.5)

$$\tilde{Y}_i = f(X, A) = \tilde{A}_0 + \tilde{A}_1 X_{i1} + \cdots + \tilde{A}_n X_{ik} \tag{2.5}$$

The function $f(X, A)$ is consider which is mapped from X into Y with the elements of X denoted by $x_i^{(t)} = (x_{i0}^t, x_{i1}^t, \ldots x_{ik}^t)$ where $t = 1, 2, \ldots, n$ and ith which represents the independent or input variables of the model. The dependent variables or response variables are denoted as Y_i in, Y where $\tilde{A} = (\tilde{A}_0, \tilde{A}_1, \ldots, \tilde{A}_n)$ are the model regression coefficients. If $\tilde{A}_j (j = 0, 1, \ldots, n)$ are given as fuzzy sets, the model $f(X, A)$ is called a fuzzy model. $\tilde{A}_j$ is fuzzy coefficient in terms of symmetric fuzzy numbers.

The fuzzy components were assumed as triangular fuzzy number (TFN). Figure 2.1 shows the fuzzy coefficient of triangular fuzzy number that flows in its center or model value, right and left spreads. Asymmetrical TFN that represented by observed data is defined by a triplet $\tilde{y}_i = (a_j, c_j^L, c_j^R)$. Therefore, if left and right are identical, the left and right functions are similar, then the TFN) is known as symmetrical TFN (STFN). The data used in this study are symmetrical triangular fuzzy numbers.

2.4 Methods to Determine Spreading Value

Because the data was a symmetrical TFNs, it is regarded as $c_j^L = c_j^R = c_j$. Figure 1.2 also shows the membership function for the symmetrical fuzzy regression coefficients, $A_j = (a_j, c_j)$. For symmetric, the response variable represented $\tilde{y}_j = (a_j, c_j)$, while a_j is the center point representing the original data value, and c_j represents the spreading value. The spreading value, c_j is obtained from [14]:

$$c_{n,j} = \frac{|y_i - \bar{y}|}{2} \tag{2.6}$$

where $Y_i (i = 1, 2, \ldots, n)$ stands for the value of WQI and $\hat{Y}$ stands for the mean of data value of WQI, Y_i. In this study, there are one fuzzy response variable and six independent variables. Thus, the fuzzy linear regression is estimated as follows:

$$\tilde{Y}_i = \beta_0 + \beta_1 X_1 + \beta_2 X_2 + \beta_3 X_3 + \beta_4 X_4 + \beta_5 X_5 + \beta_6 X_6 \qquad (2.7)$$

where $\beta_0 = (a_0, c_{0,j})$ is the fuzzy intercept coefficient, $\beta_1 = (a_1, c_{1,j})$, $\beta_2 = (a_2, c_{2,j})$, $\beta_3 = (a_3, c_{3,j})$, $\beta_4 = (a_4, c_{4,j})$, $\beta_5 = (a_5, c_{5,j})$, $\beta_6 = (a_6, c_{6,j})$ are the fuzzy slope coefficient for $X_i, i = 1, 2, 3, 4, 5, 6$. The variables $X_i, i = 1, 2, 3, 4, 5, 6$ represents the parameters of water quality where X_1 is DO, X_2 is BOD, X_3 is COD, X_4 is AN, X_5 is suspended solids and X_6 is pH.

However, in this study, we made a new improvement for determining the spreading value which it could give smaller spreading value and better results. The new method is as follows:

$$c_{n,j} = \frac{|y_i - \bar{y}|}{4} \qquad (2.8)$$

The differences between these two methods can be shown in the figures below.

By using the formula stated in (2.7), the Fig. 2.2 shows that the spreading of left and right models has a large gap compared to the Fig. 2.3 which has smaller gap and produce smaller error. Thus, we used this spreading method in our study in order to improve the model and provide a better accuracy compared with existing fuzzy based model. In Isa et al. [10], the spreading value is calculated by using formula (2.7).

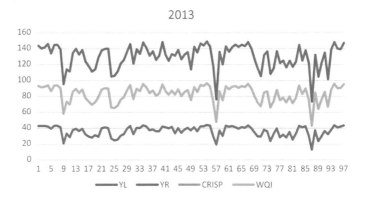

Fig. 2.2 Result for spreading value using formula (2.7)

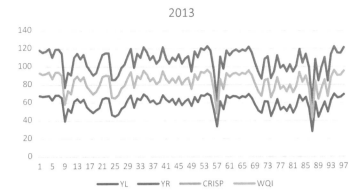

Fig. 2.3 Result for spreading value using formula (2.8)

2.5 Fuzzy Multiple Linear Regression

From [11] works, they stated that if $\tilde{Y}_i$ is a symmetrical triangular fuzzy number, $c_{i,L} = c_{i,R}$, the least square approach is used to evaluate the particular regression line $(\hat{Y}_i)$ where the sum of squared deviations above or below the $(\tilde{Y}_i)$ data points is minimized.

This method is considered symmetrical triangular membership function in the following discussion. With k crisp independent variables and one fuzzy dependent variable, the estimated fuzzy regression can be expressed as

$$\tilde{y}_i = (a_0, c_{0,j}) + (a_1, c_{1,j}) + (a_2, c_{2,j})X_2^2 + \cdots + (a_k, c_{k,j})X_k^n$$

where $(a_0, c_{0,j})$ is the fuzzy intercept coefficient; $(a_1, c_{1,j})$ is the fuzzy slope coefficient for X_1; $(a_2, c_{2,j})$ is the fuzzy slope coefficient for X_2^2; $(a_k, c_{k,j})$ is the kth fuzzy slope coefficient. The expected $\tilde{y}_i$ at the particular μ value is given by [16];

$$\mu_{\tilde{y},L} = [a_0 - (1 - \mu)c_{0,L}] + [a_1 - (1 - \mu)c_{1,L}]X_1 + \cdots + [a_k - (1 - \mu)c_{k,L}]X_k^n$$
$$= (a_0 + a_1 + \cdots + a_k) - (1 - \mu)(c_{0,L} + c_{1,L}X_1 + \cdots + c_{k,L}X_k^n)$$

and

$$\mu_{\tilde{y},R} = (a_0 + a_1 + \cdots + a_k) - (1 - \mu)(c_{0,R} + c_{1,R}X_1 + \cdots + c_{k,R}X_k^n)$$

where $a_0, a_1, \ldots, a_k$ are the expected coefficients of $\hat{y}_i$ at $\mu = 1$; $c_{0,L} + c_{1,L}X_1$ and $c_{0,R} + c_{1,R}X_1$ are the left fuzzy width and the right fuzzy width for X_1; $c_{0,L} + c_{2,L}X_2^2$ and $c_{0,R} + c_{2,R}X_2^2$ are the left fuzzy width and the right fuzzy width for X_2^2; $(1 - \mu)(c_{0,L} + c_{1,L}X_1)$ and $(1 - \mu)(c_{0,R} + c_{1,R}X_1)$ are the left fuzzy width and the right fuzzy width for X_1 at a given μ value; $(1 - \mu)(c_{0,L} + c_{2,L}X_2^2)$ and $(1 - \mu)(c_{0,R} + c_{2,R}X_2^2)$ are the left fuzzy width and the right fuzzy width for X_2^2 at a given μ value. The steps for fuzzy multiple linear regression can be expressed in the following matrix form.

$$\tilde{Y} = X\tilde{\beta} \tag{2.9}$$

where

$$\tilde{Y} = \begin{bmatrix} a_1, c_{1,j} \\ a_2, c_{2,j} \\ \vdots \\ a_n, c_{n,j} \end{bmatrix} \tag{2.10}$$

$$\hat{\tilde{\beta}} = \begin{bmatrix} \hat{\tilde{\beta}}_0 \hat{\tilde{\beta}}_1 : \hat{\tilde{\beta}}_k \end{bmatrix} = \begin{bmatrix} (a_0, (1-\mu)c_{0,j}) \\ (a_1, (1-\mu)c_{1,j}) \\ \vdots \\ (a_n, (1-\mu)c_{n,j}) \end{bmatrix} \tag{2.11}$$

and

$$X = \begin{bmatrix} 1 & x_{11} & x_{21} & \cdots & x_{k1} \\ 1 & x_{11} & x_{11} & \cdots & x_{k2} \\ \vdots & \vdots & \vdots & & \vdots \\ 1 & x_{11} & x_{11} & \cdots & x_{kn} \end{bmatrix} \tag{2.12}$$

Matrices Y and X in the above equations are the data matrices associated with the variables of response and predictor, respectively. Matrix β includes the least square estimates of the coefficients of regression. In this study, we have adopted a symmetric fuzzy triangular number as discussed in [11] and [10] as in Eqs. (2.10), (2.11) and (2.12).

To obtain the regression parameters, Eq. (2.5) can be transformed as shown in (2.13)

$$(X'X)\hat{\tilde{\beta}} = X\tilde{Y} \tag{2.13}$$

where X' is the transpose matrix of X.

The regression coefficients can be derived by matrix operations as follow:

$$\hat{\tilde{\beta}} = (X'X)^{-1}X'\tilde{Y} \tag{2.14}$$

where $(X'X)^{-1}$ is the inverse matrix of $X'X$,

$$X'X = \begin{bmatrix} n & \sum x_{1i} & \cdots & \sum x_{6i} \\ \sum x_{1i} & \sum x_{1i} \sum x_{2i} & \cdots & \sum x_{1i} \sum x_{6i} \\ \vdots & \vdots & \ddots & \vdots \\ \sum x_{6i} & \sum x_{1i} \sum x_{6i} & & \sum x_{6i}^2 \end{bmatrix} \quad (2.15)$$

$$X'Y = \begin{bmatrix} g_0 = \sum_{i=1}^{n} y_i \\ g_1 = \sum_{i=1}^{n} x_{1i} y_i \\ \vdots \\ g_k = \sum_{i=1}^{n} x_{ki} y_i \end{bmatrix} \quad (2.16)$$

2.6 Fuzzy Coefficient of Determination

The fitted fuzzy regression equation can be developed based on the estimated regression coefficients. After establishing the regression equation, it is of interest to measure the quality or reliability of the fitted regression equation. The fuzzy coefficient of determination $(HR)^2$ is used to interpret the proportion of the total variation in Y explained by the regression line, which is defined by

$$(HR)^2 = \frac{\sum_{i=1}^{n} (\hat{Y}_i - \bar{\tilde{Y}})^2}{\sum_{i=1}^{n} (\tilde{Y}_i - \bar{\tilde{Y}})^2}$$

where $\bar{\tilde{Y}}$ is the mean of fuzzy data $\tilde{Y}$.

The above expression can be represented by the following expression: $(HR)^2 =$
$$\frac{\sum_{i=1}^{n}(a_0+a_1X_i-\bar{Y})^2+(1-\mu)\sum_{i=1}^{n}(c_{0,L}+c_{1,L}X_i-\bar{e}_L)^2+(1-\mu)\sum_{i=1}^{n}(c_{0,R}+c_{1,R}X_i-\bar{e}_R)^2+(1-\mu)}{\sum_{i=1}^{n}(Y_i-\bar{Y})^2+(1-\mu)\sum_{i=1}^{n}(e_{i,L}-\bar{e}_L)^2+(1-\mu)\sum_{i=1}^{n}(e_{0,R}-\bar{e}_R)^2}$$ where, $\bar{e}_L$
and $\bar{e}_R$ are the mean of left fuzzy width and mean of right fuzzy width, respectively.

Likewise, the fuzzy correlation coefficient (HR) is the root of $(HR)^2$, which can evaluate the strength of the linear relationship between predictor variables and response variables. Next, e_i is the deviation between actual value Y_i and the estimated value $\hat{Y}_i$. It is defined as follows,

$$e_i = Y_i - \hat{Y}_i$$

There are three types that were computed which are Sum Square Regression (SSR), Sum Square Errors (SSE) and Total of Sum Square Error (SST). The equation of SSR, SSE and SST are defined as follows:

$$SSR = \sum_{i=1}^{n} (\tilde{Y}_i - \bar{\tilde{Y}}_i)$$

$$SSE = \sum_{i=1}^{n} (Y_i - \hat{Y}_i) = \sum_{i=1}^{n} e_i^2$$

$$SSR = SSR + SSE$$

Fuzzy linear regression equation, fuzzy correlation coefficient and fuzzy coefficient of determination are the main measurement that were obtained from water quality index (WQI). The higher the value of $(HR)^2$, the better the fuzzy regression model.

References

1. Abdullah L, Zakaria N (2012) Matrix driven multivariate fuzzy linear regression model in car sales. J Appl Sci (Faisalabad) 12(1):56–63
2. Anand MCJ, Bharatraj J (2017) Theory of triangular fuzzy number. Proc NCATM 2017:80
3. Asai HTSUK, Tanaka S, Uegima K (1982) Linear regression analysis with fuzzy model. IEEE Trans. Systems Man Cybern 12:903–907
4. Bhavyashree S, Mishra M, Girisha GC (2017) Fuzzy regression and multiple linear regression models for predicting mulberry leaf yield: a comparative study. Int J Agric Stat Sci 13(1):149–152
5. Buckley JJ, Eslami E (2002) An introduction to fuzzy logic and fuzzy sets, vol 13. Springer Science & Business Media, New York
6. Chang PT, Lee ES (1996) A generalized fuzzy weighted least-squares regression. Fuzzy Sets Syst 82(3):289–298
7. Chang YHO, Ayyub BM (1997) Hybrid fuzzy regression analysis and its applications. In: Uncertainty modeling and analysis in civil engineering, pp 33–41
8. D'Urso P, Gastaldi T (2001) Linear fuzzy regression analysis with asymmetric spreads. In: Advances in classification and data analysis. Springer, Berlin, Heidelberg, pp 257–264
9. Hidayah Mohamed Isa N, Othman M, Karim SAA (2018) Multivariate matrix for fuzzy linear regression model to analyse the taxation in Malaysia. Int J Eng Technol 7(4.33):78–82
10. Lowen R (1976) Fuzzy topological spaces and fuzzy compactness. J Math Anal Appl 56(3):621–633
11. Lowen R (1996) Fuzzy real numbers. In: Fuzzy set theory. Springer, Dordrecht, pp 143–168
12. Pan NF, Lin TC, Pan NH (2009) Estimating bridge performance based on a matrix-driven fuzzy linear regression model. Autom Constr 18(5):578–586
13. Rommelfanger H, Słowiński R (1998) Fuzzy linear programming with single or multiple objective functions. In: Fuzzy sets in decision analysis, operations research and statistics. Springer, Boston, MA, pp 179–213
14. Sakawa M, Yano H (1992) Multiobjective fuzzy linear regression analysis for fuzzy input-output data. Fuzzy Sets Syst 47(2):173–181
15. Tranmer M, Elliot M (2008) Multiple linear regression. Cathie Marsh Cent Census Surv Res (CCSR) 5:30–35

16. Ubale A, Sananse S (2016) A comparative study of fuzzy multiple regression model and least square method. Int J Appl Res 2(7):11–15
17. Voskoglou M (2015) Use of the triangular fuzzy numbers for student assessment. arXiv preprint arXiv:1507.03257
18. Zadeh LA (1965) Fuzzy sets. Inf Control 8(3):338–353

Chapter 3
Water Quality Index (WQI)

Abstract In this chapter, we discuss the formulation of WQI that have been used by DOE, Malaysia to measure the quality of the river as well as drinking water in Malaysia. We begin with the elaboration on all six parameters that will used in the formulation later. These parameters values will be measured at each river and will be embedded into the formulation to determine WQI value that will classify the class of the water.

Keywords Water quality index (WQI) · Ecosystem · Parameters · Water qualification · Interim water quality standard

3.1 Introduction

Rivers are vital to human and animals as they are main sources of water to all wellbeing and play an important role in balancing the ecosystem. Rivers are also vulnerable to pollution as rapid development has produced tremendous amounts of human wastes, including domestic, industrial, commercial and transportation wastes which inevitably ends up in the water bodies. Hence, the authority (The Department of Environment (DOE)) took a new initiative and comes up with River Water Quality Monitoring Program to determine the river water quality status and detect changes over time [1]. The program has been implemented since 1978 and widely used starting in 1995. In order to evaluate the status of river water quality, DOE uses Water Quality Index (WQI) and classify the rivers based on the designation of classes beneficial uses provided by National Water Quality Standard for Malaysia (NWQS) [16].

3.2 Water Quality Characteristics

As humans progress in transportation and manufacturing, our environment is also at stake to pollution. All of those activities produce a lot of wastages and pollute our air, water and soil. However, by identifying water quality parameters and the level of pollutants, we can determine whether the water is suitable for certain usage or

S. A. A. Karim and N. F. Kamsani, *Water Quality Index Prediction Using Multiple Linear Fuzzy Regression Model*, SpringerBriefs in Water Science and Technology,
https://doi.org/10.1007/978-981-15-3485-0_3

not. There are many water quality parameters that affect the quality of water in the environment. They can be divided into three properties or characteristics which are biological, chemical and physical.

The examples of physical properties are temperature, salinity and turbidity which can be determined by four senses of touch, sight, smell and taste. For a clearer explanation, we can say that we can sense the temperature of the water by touch, turbidity and suspended solids by sight, and odour by smell. The water quality is significantly affected by each of these parameters [6].

Each of most aquatic species is poikilothermic—i.e., "cold-blooded"—meaning that they are unable to control their core body temperature internally [13]. Different species of aquatic organisms have different temperature ranges to live and there is a very few of them that can tolerate with high changes in temperature. Therefore, temperature holds a very crucial part in balancing water ecosystem.

Turbidity is referred to how transparent and clear the water is. A high number of individual particles or total suspended solid (TSS) can cause the water to lose its transparency and damage the fish habitats and other aquatic organisms [18].

Total suspended solids is usually related with turbidity and referred to particles in water which is usually larger than 0.45 μm [6]. Suspended solids can cause the water colour to darken, decreasing the quantity of light available to grow by photosynthesis for aquatic plants, algae and mosses.

Chemical properties include parameters such as pH (potential of hydrogen), nitrate, dissolved oxygen (DO), biochemical oxygen demand (BOD), chemical oxygen demand (COD), metals and oil. pH parameter is used to measure the amount of acid in the water. The pH scale runs from the range of 0 (very acidic) to 14 (very alkaline) with pH 7 is the neutral condition in the water [8]. The natural pH for most water is usually range between 6.0 and 8.5 [5].

Like other living things, fish and aquatic organisms need oxygen to live. The oxygen content of natural waters varies with temperature, salinity, turbulence, the photosynthetic activity of algae and plants and atmospheric pressure. Natural water's oxygen content differs with temperature, salinity, turbulence, algae and plant photosynthesis activity, and atmospheric pressure [8]. Dissolved oxygen (DO) at sea level ranges between 15 mg/L at 0 °C and 8 mg/L at 25 °C in freshwaters. Concentrations are generally close to, but less than, 10 mg/L in unpolluted waters [5].

Biochemical oxygen demand (BOD) is used to determine the measure of organic pollution resulting from agro-based and other industries to stabilize domestic and industrial waste. BOD also measures the organic matter that serves as food (mainly organic) for the bacteria in the water. BOD has been used for more than a century to estimate the amount of biodegradable waste is present in the water [21]. COD test is used to measure chemical pollution in the water. Scientifically, it is used to determine the number of organic and inorganic oxidizable compounds in the water.

From DOE, ammoniacal nitrogen is used to measure organic pollution from sewage and animal waste; sewage (human and livestock). Ammonia nitrogen is frequently produced by product breakdowns of urea and protein, so it is usually abundant as domestic waste in the hydrological ecosystem [19]. Human activity is well known to play a significant part in the lake ecosystems as a contributor to the abundance of ammonia nitrogen in which levels may exceed 5 mgL^{-1} [3].

Biological parameters should also be considered in order to evaluate the water quality. Usually biological parameters involve pathogens and indicator organisms. In biology, a pathogen is anything that can cause disease in the earliest and broadest sense. They reach primarily water bodies from untreated sewage discharges. The examples of indicator organisms are faecal coliform, faecal streptococci and E-coli, used to show the possibility of sewage pollution in the water. The greater the concentration of coliform bacteria, the greater the possibility of having water pathogens. Instead, indicator organisms are used because of the difficulty in diagnosing pathogens.

For WQI calculation, there are six parameters need to be considered which are ammoniacal nitrogen (AN), biochemical oxygen demand (BOD), chemical oxygen demand (COD), dissolved oxygen (DO), pH and total suspended solid (SS). Further discussion on WQI calculation and standard formula will be shown in the next section.

3.3 Water Quality Index

Since long time ago, water has been frequently used for many purposes in humans' daily life. In order to determine status changes of water quality, DOE has developed Water Quality Index (WQI) and used National Water Quality Standard for Malaysia (NWQS) as a tool for guidelines in monitoring and maintaining Malaysia's river water quality. WQI has been used as a mathematical tool which aggregates data on two or more water quality parameters to produce a single number [17]. On a scale from 0 to 100, the WQI can be specified as a number used to display the water quality. It consists dissolved oxygen (DO), total phosphorus and fecal coliform which has specific impacts to beneficial uses.

There are several water quality indices that have been developed since 1965 to assess water bodies. Most of these indices, however, are based on the U.S. developed WQI. Foundation of National Healthcare (NSF) [18]. The current method of estimating the WQI in Malaysia is based on opinion polls [12].

3.3.1 National Sanitation Foundation Water Quality Index

Initially in 1970, NSF WQI has been developed by National Sanitation Foundation (NSF) [2]. About 142 persons with known expertise in water quality management was assembled throughout the U.S.A to conduct a survey. The survey was conducted to conclude which water quality tests out of about 35 tests should be included in an index. It was intended to include nine parameters to conduct an integrated evaluation of the circumstances of water quality in order to fulfill utilization goals. The index is computed as the weighted sum of sub-indices where the sum of the sub-indices

Table 3.1 NSF WQI
analytes and weights

Parameter/analyte	WQI weights
Dissolved oxygen (% sat)	0.17
Fecal coliform (or E.coli) (#/100 mL)	0.16
pH (standard units)	0.11
Biochemical oxygen demand (mg/L)	0.11
Temperature change (°C)	0.10
Total phosphate (mg/L)	0.10
Nitrates (mg/L)	0.10
Turbidity (NTU)	0.08
Total suspended solids (mg/L)	0.07

Table 3.2 NSF WQI quality
scale [21]

WQI	Quality of water
91–100	Excellent
71–90	Good
51–70	Medium or average
26–50	Fair
0–25	Poor

is equal to 1. A rating curve for each of the nine parameters gives different level of water quality from 0 to 100. The y-axis indicates levels of water quality from 0 to 100, while x-axis indicates increasing levels of the particular analyte. The nine parameters and their corresponding weights are listed in Table 3.1 [8]. The water quality value was then determined as the product of the value of the rating curve and the weight of the WQI [21].

Once the overall WQI score is determined, to decide how good the water is on a given day, it can be compared to a scale given in Table 3.2.

3.3.2   Standard Water Quality Index

The water quality index developed by the Environment Department (DOE) has been used for about 25 years in Malaysia [9]. There are several steps or processes in calculating WQI. The three main steps are summarized below:

1. Collect water sample at various location and basin.
2. Measurement of the six parameters whether by using in situ measurements or laboratory analysis of collected sample.
3. Apply the WQI formula to determine the class of the river.

There are six water quality parameters required to measure WQI such as dissolved oxygen (DO), biological oxygen demand (BOD), chemical oxygen demand (COD), suspended solid (SS), the pH value (pH), and ammoniacal nitrogen (NH$_3$–NL) [12]. Among all six parameters, DO carries maximum weightage of 0.22 and pH carries the minimum of 0.12 in the WQI equation. The formula used in the calculation of WQI is as defined as follows:

$$WQI = 0.22 * SIDO + 0.19 * SIBOD + 0.16 * SICOD + 0.15 * SIAN$$
$$+ 0.16 * SISS + 0.12 * SIpH \tag{3.1}$$

where;

SIDO   Subindex DO (% saturation)
SIBOD   Subindex BOD
SICOD   Subindex COD
SIAN   Subindex NH$_3$–NL
SISS   Subindex SS
SIpH   Subindex pH.

$$0 \le WQI \le 100$$

Sub-index for DO (in % saturation):

$$SI_{DO} = 0 \qquad\qquad\qquad\qquad\qquad \text{for DO} < 8$$
$$= 100 \qquad\qquad\qquad\qquad\qquad \text{for DO} > 92$$
$$= -0.395 + 0.030 \, DO^2 - 0.00020DO^3 \quad \text{for } 8 < DO < 92 \tag{3.2}$$

Sub-index for BOD:

$$SI_{BOD} = 100.4 - 4.23BOD \qquad\qquad \text{for BOD} < 5$$
$$= 108e^{-0.055BOD} - 0.1BOD \quad \text{for BOD} > 5 \tag{3.3}$$

Sub-index for COD:

$$SI_{COD} = -1.33COD + 99.1 \qquad\qquad \text{for COD} < 20$$
$$= -103e^{-0.0157COD} - 0.04COD \quad \text{for COD} > 20 \tag{3.4}$$

Sub-index for AN:

$$SI_{AN} = 100.5 - 105AN \qquad\qquad \text{for AN} < 0.3$$
$$= 94e^{-0.573AN} - 5|AN - 2| \qquad \text{for } 0.3 < AN < 4$$
$$= 0 \qquad\qquad\qquad\qquad\qquad \text{for AN} > 4 \tag{5.5}$$

Sub-index for SS:

$$\begin{aligned}
SI_{SS} &= 97.5e^{-0.00676SS} + 0.05SS & \text{for } SS < 100 \\
&= 71e^{-0.0016SS} - 0.015SS & \text{for } 100 < SS < 1000 \\
&= 0 & \text{for } SS > 1000
\end{aligned} \tag{5.6}$$

Sub-index for pH:

$$\begin{aligned}
SI_{pH} &= 17.2 - 17.2pH + 5.02pH^2 & \text{for } pH < 5.5 \\
&= -242 + 95.5pH - 6.67pH^2 & \text{for } 5.5 < pH < 7 \\
&= -181 + 82.4pH - 6.05pH^2 & \text{for } 7 < pH < 8.75 \\
&= 536 - 77.0pH + 2.76pH^2 & \text{for } pH > 8.75
\end{aligned} \tag{6.7}$$

100 is the highest possible score and denotes a pristine river and zero is the lowest. The WQI score can then be used to categorize a particular water body into one of five classes. Based on [7], water qualification can be classified into five classes which is shown in the Table 3.3.

Advanced water quality and ecologically based standards that integrate physical, chemical and biological numerical criteria provide the potential for better understanding, management, conservation and restoration of water bodies [14] The Interim National Water Quality Standard, Malaysia (INWQS) categorizes surface water in Malaysia into six classes as stated in Table 3.4.

This chapter discusses the definition of WQI that have been used by DOE, Malaysia in calculating the water quality and to classify it into one of five water

Table 3.3 Water qualification based on parameters

Parameter	Unit	Class				
		I	II	III	IV	V
Ammoniacal nitrogen	mg/L	<0.1	0.1–0.3	0.3–0.9	0.9–2.7	>2.7
Biochemical oxygen demand	mg/L	<1	1–3	3–6	6–12	>12
Chemical oxygen demand	mg/L	<10	10–25	25–50	50–100	>100
Dissolved oxygen	mg/L	>7	5–7	3–5	1–3	<1
pH		>7.0	6.0–7.0	5.0–6.0	<5.0	>5.0
Total suspended solid	mg/L	<25	25–50	50–150	150–300	>300
WQI		>92.7	76.5–92.7	51.9–76.5	31.0–51.9	<31
Pollution degree		Very clean	Clean	Moderate	Slightly polluted	Polluted

Table 3.4 Water quality classes and uses (Interim national water quality standards for Malaysia)

Class I	Conservation of natural environment Water supply I—practically no treatment necessary Fishery I—very sensitive aquatic species
Class IIA	Water supply II—conventional treatment required Fishery II—sensitive aquatic species
Class IIB	Recreational use with body contact
Class III	Water supply III—extensive treatment required Fishery III—common, of economic value and tolerant species; livestock drinking
Class IV	Irrigation
Class V	None of the above

classes. Water with class I or II is safe for us to drink and for other purposes. Thus, it is important for the relevant government agency to provide us the correct water quality for each river as well as other water sources such as tap water, drinking water, mineral water etc.

References

1. ASMA (2012) River water quality monitoring. http://www.doe.gov.my/portalv1/en/general-info/pemantauan-kualiti-air-sungai/280. Accessed from 13 May 2019
2. Brown RM, McClelland NI, Deininger RA, Tozer RG (1970) A water quality index- do we dare
3. Cane F, Hoxha B, Avdolli M (2010) Water quality in Carstic lake, Albania. Nat Monten Podgor 9(3):349–355
4. Canter LW (2018) Environmental impact of water resource projects. CRC Press
5. Chapman D, Kimstach V (1996) Selection of water quality variables (Chap 3). In: Water quality assessments-a guide to use of biota, sediments and water in environmental monitoring, 2nd ed. Published on behalf of WHO by F & FN Spon
6. DID (Department of Irrigation and Drainage) (2009) Study on the river water quality trends and indexes in Peninsular Malaysia. Department of Irrigation and Drainage, Ministry of Agriculture, Kuala Lumpur
7. Department of Environment Malaysia (2005) Interim national water quality standards for Malaysia. http://www.doe.gov.my/index.php?option=com.content&task=view&id=244&Itemid=615&lang=en. Accessed from 17 January 2008
8. Elshemy MM (2010) Water quality modeling of large reservoirs in semi-arid regions under climate change-example Lake Nasser (Egypt). Doctoral dissertation, Technische Universität Braunschweig
9. Huang YF, Ang SY, Lee KM, Lee TS (2015) Quality of water resources in Malaysia. Res Pract Water Qual
10. Int J Res Eng Technol 2(4):609–614
11. Khalik WMAWA, Abdullah MP (2012) Seasonal influence on water quality status of Temenggor Lake, Perak. Malays J Anal Sci 16:163–171
12. Khuan LY, Hamzah N, Jailani R (2002) Prediction of water quality index (WQI) based on artificial neural network (ANN). In: Student conference on research and development. IEEE, pp 157–161

13. Lakesuperiorstreams (2009) LakeSuperiorStreams: community partnerships for understanding water quality and storm water impacts at the head of the great lakes. http://lakesuperiorstreams.org. University of Minnesota-Duluth, Duluth, MN 55812
14. Magner JA, Brooks KN (2008) Integrating sentinel watershed-systems into the monitoring and assessment of Minnesota's (USA) waters quality. Environ Monit Assess 138(1–3):149–158
15. Michaud JP (1991) A citizen's guide to understanding and monitoring lakes and streams. Envirovision-Environmental Consulting Service
16. Nurul-Ruhayu MR, An YJ, Khairun Y (2015) Detection of river pollution using water quality index: a case study of tropical rivers in Penang Island, Malaysia. Open Access Libr J 2(03):1
17. Ott W (1978) Water quality indices: a survey of indices used in the United States, vol 1. Environmental Protection Agency, Office of Research and Development, Office of Monitoring and Technical Support
18. Said A, Stevens DK, Sehlke G (2004) An innovative index for evaluating water quality in streams. Environ Manag 34(3):406–414
19. Shafiul C, Karen G (2006) The impact of land use on surface water quality in Queens County, New York. J Environ Hydrol 14(15):1–7
20. Srivastava G, Kumar P (2013) Water quality index with missing parameters
21. Washington State Department of Ecology (2002) Introduction to water quality index. http://www.fotsch.org/WQI.htm. Accessed 20 January 2008

Chapter 4
Data Collection and Study Sites

Abstract This chapter discuss the details of the data collection and study sites of the research.

4.1 Perak River Location

The case study is situated in the state of Perak, Malaysia, where it covers an area of 21,035 km^2, making up 6.4% of total land mass in Malaysia. It is the second-largest Malaysian state in the Malaysia Peninsula, and the fourth-largest in the whole of Malaysia. The latitude and longitude coordinates of Perak are: 4.597479° N, 101.090103° E. There are 11 major river basins that covers over 80 km^2. Perak river basin is the largest basin with an area of 14.908 km^2 where it covers about 70% of the total area of the state area and the length is 400 km. It is also the second largest river basin in Peninsular Malaysia after Pahang river basin. The source of Perak River is from the mountainous Perak-Kelantan-Thailand border of the Belum Forest Reserve.

In this study, we chose 8 river basins which are Kurau, Sepetang, Bruas, Perak, Raja Hitam, Bernam, Wangi and Kerian with 24 rivers as shown in the Table 4.1.

4.2 Data Collection

The sample and data collection were duly carried out by the Department of Environment (DOE) of Malaysia through Alam Sekitar Malaysia Sdn. Bhd. [1]. The duration of the data is from year 2013 until year 2017 consisting all rivers in Sungai Perak i.e. about more than 1110×17 data sets. The data consist of 8 river basins, 27 rivers, station numbers and 9 parameters of water quality. The samples are taken at selected point or location using an automatic sampling approach as well as manually taken once a week or twice a month.

S. A. A. Karim and N. F. Kamsani, *Water Quality Index Prediction Using Multiple Linear Fuzzy Regression Model*, SpringerBriefs in Water Science and Technology, https://doi.org/10.1007/978-981-15-3485-0_4

31

Table 4.1 The location of the study area

States	Basin	STA no	River
Perak	Kurau	2KU07	Air Hitam
		2KU01	Ara
		2KU02	Kurau
	Sepetang	2SP08	Batu Tegoh
		2SP01	Sepetang
		2SP11	Temerloh
		2SP16	Trong
	Bruas	2BR01	Bruas
		2BR06	Dandang
		2BR02	Rotan
	Perak	2PK02	Bidor
		2PK31	Kampar
		2PK19	Kinta
		2PK62	Manong
		2PK54	Nyamok
		2PK01	Perak
	Raja Hitam	2RH01	Manjong
		2RH11	Nyior
		2RH03	Raja Hitam
	Bernam	1BM15	Gelinting
		1BM09	Trolak
	Wangi	2WD07	Deralik
		2WD09	Wangi
	Kerian	2KR09	Selama

If water in Perak river basin is contaminated, it will affect most of the river in Perak state and most likely it will affect the locals where their income depend on the fishing and agriculture activities in that area. The aim of this study is to predict the water quality of the Perak river basin in real time. If the river water quality is predicted as poor or polluted, some preventive measures can be taken immediately. Each parameter was analyzed based on the Water Quality Standard and Regulation in Malaysia.

Among that information are the six variables that are taken as the independent variables for this study. They are:

1. SI-Sub index of parameter
2. DO-Dissolve Oxygen
3. BOD-Biological Oxygen Demand
4. COD-Chemical Oxygen Demand
5. AN-Ammonical Nitrogen

6. TSS-Suspended Solid
7. pH-Salinity.

All of these data are quantitative in nature. For 2013, there are overall 213 data from DOE but we only analyzed about 98 data from it. It goes the same for the data from 2014 to 2017. The data are being processed and filtered first with some cleaning method including data classification from clean river to bad river etc. Water quality data are used to assess the status by analyzing whether it is clean, slightly polluted, or contaminated. Then, the proposed method is applied. Figures 4.1, 4.2, 4.3 and 4.4 show the map of major river basins in Perak state.

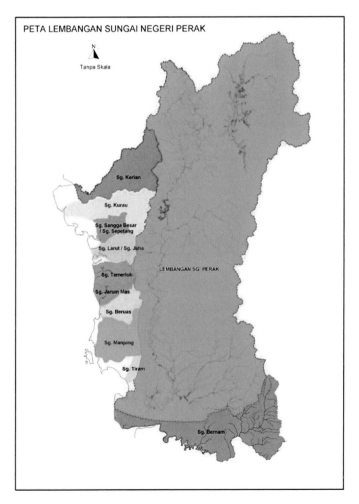

Fig. 4.1 Map of major basins in Perak state [2]

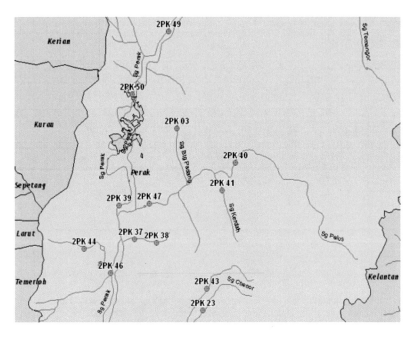

Fig. 4.2   Perak (North) river basin monitoring stations [1]

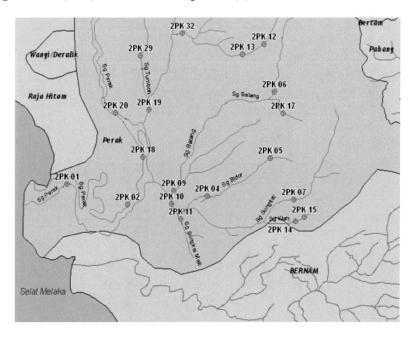

Fig. 4.3   Perak (South) river basin monitoring stations [1]

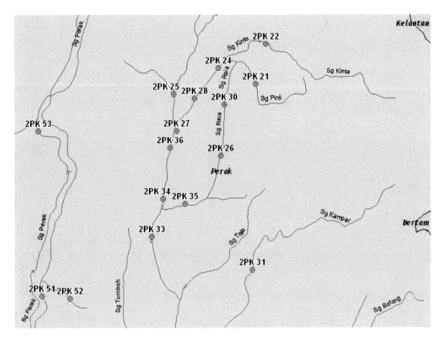

Fig. 4.4 Perak (Central) river basin monitoring stations [1]

References

1. ASMA (2012) River water quality monitoring. http://www.doe.gov.my/portalv1/en/general-info/pemantauan-kualiti-air-sungai/280. Accessed from 13 May 2019
2. DID (Department of Irrigation and Drainage) (2018) The river, basins and reserves. Accessed from http://jps.perak.gov.my/index.php/en/about-us-en/section/river-section/general-information/77-about-us/275-river-basin-reserve

Chapter 5
Water Quality Index Using Fuzzy Regression

Abstract In this chapter, results and findings from the models, techniques and algorithms developed in the previous chapters are presented and discussed in details. The main emphasis is on the decision models of river WQI. The experiments and testing were conducted using the actual data of WQI from every river basin located in Kurau, Sepetang, Bruas, Perak, Raja Hitam, Bernam, Wangi and Kerian, which have been proposed and prepared in the fuzzy regression model.

Keywords Convex combination · Error analysis · Fuzzy coefficient of determination · Coefficient of correlation · SSR · SST · SSE

5.1 WQI Calculation

Based on the calculation of the spreading value using formula in (3.4), the snapshot of the data from year 2013 before preprocessing can be shown in the Table 5.1.

Now, the WQI value for the rivers located in Perak can be calculated by using the Eqs. (2.10)–(2.12) in Chap. 2. The result for WQI coefficients are shown below:

Table 5.1 Fuzzy multivariate regression and spreading value based on formula

Theoretical value, Y_i	Spreading value, $C_{n,j}$
93.35907	2.52747
91.69805	2.112213
92.5241	2.318727
94.21343	2.741058
86.88096	0.907941
94.09605	2.711713
94.1512	2.725502
90.45846	1.802316
58.44642	6.200694
73.8078	2.360348

(continued)

Table 5.1 (continued)

Theoretical value, Y_i	Spreading value, $C_{n,j}$
70.21911	3.25752
86.1526	0.725852
89.74366	1.623617
84.70996	0.36519
88.91646	1.416817
78.33365	1.228885
73.93473	2.328616
69.90271	3.336621
72.82537	2.605956
79.35934	0.972464
88.93775	1.422139
90.56516	1.828991
90.16991	1.730179
66.19661	4.263147
65.45582	4.448344
68.68837	3.640206
76.19531	1.763472
79.35845	0.972687
87.91702	1.166955
94.35493	2.776433
77.03633	1.553217
89.96871	1.679879
86.91922	0.917505
95.84654	3.149335
91.67933	2.107533
84.41409	0.291223
87.583	1.083451
81.12815	0.530261
83.91901	0.167453
95.23365	2.996115
86.61483	0.841409
82.34957	0.224907
87.45727	1.052018
82.43976	0.202359
89.32556	1.519092
80.48246	0.691683
85.58762	0.584607
88.10433	1.213784

(continued)

Table 5.1 (continued)

Theoretical value, Y_i	Spreading value, $C_{n,j}$
75.62075	1.90711
91.88461	2.158853
85.73709	0.621973
94.47046	2.805315
93.26052	2.50283
96.59758	3.337097
92.86912	2.404982
74.0627	2.296625
48.03675	8.80311
86.57485	0.831413
75.30562	1.985895
93.00164	2.438111
89.05046	1.450315
92.35131	2.275529
93.17106	2.480467
90.89248	1.910822
93.18338	2.483545
91.86713	2.154483
95.92302	3.168455
90.28865	1.759864
80.99153	0.564416
72.58961	2.664897
67.62004	3.907288
85.08174	0.458137
86.3413	0.773026
66.10101	4.287045
73.96043	2.322192
88.24447	1.248819
75.35327	1.973981
78.45672	1.198119
72.48236	2.691709
80.09117	0.789505
71.47928	2.94248
78.38131	1.21697
93.91245	2.665814
83.12356	0.031408
90.17376	1.731141
76.42491	1.706071

(continued)

Table 5.1 (continued)

Theoretical value, Y_i	Spreading value, $C_{n,j}$
43.30446	9.986183
84.64627	0.349268
64.33713	4.728016
76.08043	1.792192
85.8029	0.638426
67.15563	4.023391
88.42387	1.293669
96.07294	3.205937
90.59194	1.835687
90.74354	1.873587
95.39284	3.035912

$$\hat{\beta} = \begin{bmatrix} 1.322542 & 0.001071 & 0.004623 & -0.008357 & -0.00281 & -0.00271 \\ 0.001071 & 3.11E{-}05 & 1.38E{-}06 & -1.17E{-}05 & -1.9E{-}05 & -3.5E{-}06 \\ 0.004623 & 1.38E{-}06 & 0.001053 & -0.001107 & -4.5E{-}05 & 1.73E{-}05 \\ -0.00836 & -1.2E{-}05 & -0.00111 & 0.001253 & 3.42E{-}05 & -3.2E{-}05 \\ -0.00281 & -1.9E{-}05 & -4.5E{-}05 & 3.42E{-}05 & 5.35E{-}05 & 6.35E{-}06 \\ -0.00271 & -3.5E{-}06 & 1.73E{-}05 & -3.16E{-}05 & 6.35E{-}06 & 3.37E{-}05 \\ -0.00704 & -1.1E{-}05 & 7.14E{-}06 & -1.04E{-}05 & 2.84E{-}06 & 7.33E{-}06 \end{bmatrix}$$

$$\begin{bmatrix} -0.00704 \\ -1.1E{-}05 \\ 7.14E{-}06 \\ -1E{-}05 \\ 2.84E{-}06 \\ 7.33E{-}06 \\ 7.95E{-}05 \end{bmatrix}^{-1} \times \begin{bmatrix} 8075.172 & 204.79999 \\ 697695.1 & 15475.1 \\ 647655.7 & 15274.043 \\ 638142.5 & 15186.798 \\ 684583.7 & 16553.607 \\ 698175.9 & 16961.281 \\ 749096.3 & 8663.144 \end{bmatrix}$$

$$\hat{\beta} = \begin{bmatrix} -1.18234E{-}11 & 7.28625 \\ 0.22 & -0.03182 \\ 0.19 & -0.0849 \\ 0.16 & 0.05491 \\ 0.15 & 0.014337 \\ 0.16 & -0.00942 \\ 0.12 & -0.00502 \end{bmatrix}$$

The proposed generic regression model (WQI) is to predict other extreme river. Table 5.2 shows the center and the spreading value for each β.

Table 5.2 Fuzzy multivariate regression coefficient

	β_0	β_1	β_2	β_3	β_4	β_5	β_6
a_j	$-1.1823E{-}11$	0.22	0.19	0.16	0.15	0.16	0.12
c_j	7.2863	-0.0318	-0.0849	0.0549	0.01434	-0.0094	-0.0050

Fuzzy multivariate regression form for membership function, $\mu = 0$;

$$\tilde{Y}_i = (-1.18234E{-}11, 7.2863) + (0.22, -0.0318)DO + (0.19, -0.0849)BOD$$
$$+ \ (0.16, 0.0549)COD + (0.15, 0.0143)AN + (0.16, -0.0094)SS$$
$$+ \ (0.12, -0.0050)pH$$

Fuzzy multivariate regression can also be written as follows:

$$\tilde{Y}_L = -7.28625 + 0.1882DO + 0.1051BOD + 0.1051COD$$
$$+ \ 0.1357AN + 0.1506SS + 0.1149pH$$

$$\tilde{Y}_R = 7.28625 + 0.2518DO + 0.2749BOD + 0.2149COD$$
$$+ \ 0.1643AN + 0.1694SS + 0.1250pH$$

where;

$\tilde{Y}_L$ Left fuzzy multivariate regression
$\tilde{Y}_R$ Right fuzzy multivariate regression

and the crisp equation i.e. the standard multiple linear fitting is given as:

$$\tilde{Y}_i = -1.18234E{-}11 + 0.22DO + 0.19BOD + 0.16COD$$
$$+ \ 0.15AN + 0.16SS + 0.12pH$$

The result of WQI value provided by all three fuzzy models are shown in Table 5.3.

To evaluate the relationship between response variable and independent variable, we analyzed three type of errors which are Sum Square Regression (SSR), Sum Square Errors (SSE) and Total Sum Square (SST). The error analysis is also used as a result of modelling and predicting. The estimated equation is further investigated using error analysis to determine the stability of the model [1, 2, 4, 5].

The SSR for response variable, Y and errors are computed and obtained as:

- For Y, SSR $= 10686.00296$
- For Y_L, SSR $= 59974.6456$
- For Y_R, SSR $= 68737.97836$.

Table 5.3 Result of WQI with all three fuzzy models

YL	Y	YR	WQI
68.14746	93.35907	116.6317	93.35907
67.36395	91.69805	114.0368	91.69805
67.74712	92.5241	115.3292	92.5241
68.21507	94.21343	118.9156	94.21343
63.18947	86.88096	108.2559	86.88096
68.66704	94.09605	117.9874	94.09605
68.7192	94.1512	118.2166	94.1512
65.90993	90.45846	112.2892	90.45846
39.78209	58.44642	92.13757	58.44642
53.63403	73.8078	96.45896	73.8078
49.50764	70.21911	94.88336	70.21911
61.92141	86.1526	107.599	86.1526
64.6041	89.74366	111.6122	89.74366
60.67276	84.70996	107.2085	84.70996
64.12111	88.91646	112.6653	88.91646
55.39635	78.33365	100.6424	78.33365
51.81207	73.93473	98.77325	73.93473
49.09085	69.90271	93.10472	69.90271
52.11602	72.82537	94.33868	72.82537
54.37596	79.35934	101.8877	79.35934
64.46366	88.93775	110.4298	88.93775
65.86799	90.56516	112.8144	90.56516
65.20966	90.16991	113.0215	90.16991
46.68824	66.19661	84.4645	66.19661
45.11371	65.45582	87.52405	65.45582
47.24738	68.68837	90.62388	68.68837
53.51371	76.19531	95.30787	76.19531
55.99713	79.35845	98.98622	79.35845
63.72057	87.91702	109.2129	87.91702
68.59773	94.35493	118.5731	94.35493
54.82341	77.03633	95.434	77.03633
65.27896	89.96871	113.2501	89.96871
63.56032	86.91922	106.887	86.91922
69.75067	95.84654	120.8855	95.84654
67.01283	91.67933	114.0686	91.67933
60.97671	84.41409	107.3801	84.41409
62.9964	87.583	112.8307	87.583

(continued)

Table 5.3 (continued)

YL	Y	YR	WQI
58.67974	81.12815	105.3935	81.12815
60.0683	83.91901	107.9463	83.91901
68.70854	95.23365	120.2376	95.23365
63.93501	86.61483	106.1773	86.61483
61.10303	82.34957	100.0004	82.34957
64.57048	87.45727	107.4259	87.45727
57.97387	82.43976	111.2838	82.43976
64.69491	89.32556	111.4655	89.32556
57.38743	80.48246	107.6351	80.48246
61.98607	85.58762	109.4716	85.58762
64.34616	88.10433	108.5947	88.10433
56.43513	75.62075	89.4867	75.62075
66.59976	91.88461	117.2052	91.88461
60.91748	85.73709	112.5486	85.73709
68.40863	94.47046	118.8954	94.47046
67.83292	93.26052	116.6908	93.26052
70.35166	96.59758	122.1009	96.59758
68.00973	92.86912	115.942	92.86912
51.95119	74.0627	96.47009	74.0627
33.77565	48.03675	72.68989	48.03675
61.83203	86.57485	109.9556	86.57485
52.78637	75.30562	98.63712	75.30562
68.07105	93.00164	116.3892	93.00164
65.45635	89.05046	110.7668	89.05046
67.56131	92.35131	115.0909	92.35131
67.19714	93.17106	117.4604	93.17106
65.179	90.89248	115.7805	90.89248
67.29912	93.18338	118.7515	93.18338
66.17564	91.86713	114.933	91.86713
69.72938	95.92302	121.3167	95.92302
65.18055	90.28865	112.8631	90.28865
57.8756	80.99153	107.4524	80.99153
51.19711	72.58961	110.2518	72.58961
48.59913	67.62004	82.64484	67.62004
61.56202	85.08174	105.8706	85.08174
61.28784	86.3413	110.6399	86.3413
44.95885	66.10101	98.26804	66.10101

(continued)

Table 5.3 (continued)

YL	Y	YR	WQI
53.22504	73.96043	92.03407	73.96043
63.98895	88.24447	110.1822	88.24447
52.00471	75.35327	105.7377	75.35327
55.18374	78.45672	107.4857	78.45672
50.44957	72.48236	106.7824	72.48236
57.66709	80.09117	100.0697	80.09117
48.58646	71.47928	110.3516	71.47928
55.68857	78.38131	104.1568	78.38131
68.45841	93.91245	117.6924	93.91245
62.12792	83.12356	100.2487	83.12356
66.30856	90.17376	111.4985	90.17376
53.68972	76.42491	98.81382	76.42491
28.19537	43.30446	65.93432	43.30446
60.89816	84.64627	107.7319	84.64627
44.02375	64.33713	83.89331	64.33713
52.91765	76.08043	100.401	76.08043
61.18305	85.8029	107.416	85.8029
49.82596	67.15563	80.47815	67.15563
63.15313	88.42387	115.2293	88.42387
69.91452	96.07294	121.3897	96.07294
65.7628	90.59194	112.8657	90.59194
66.4094	90.74354	112.8151	90.74354
69.45654	95.39284	120.1156	95.39284

The SSE is computed with the similar fashion. It is found that:

- For Y, SSE $= 1.3562E-20$
- For Y_L, SSE $= 53669.8549$
- For Y_R, SSE $= 53669.8549$.

The SST is computed using the relation SSR $+$ SSE $=$ SST which is shown below:

- For Y, SST $= 113644.5005$
- For Y_L, SST $= 122407.8333$
- For Y_R, SST $= 10686.00296$.

The error analysis can be used to validate the model based on correlation measure. This correlation is termed as coefficient of determination $(HR)^2$ aiming to see the contribution of predictors to the overall WQI.

Table 5.4 Interpretation of correlation coefficients [3]

Range of correlation coefficients	Degree of correlation
0.8–1.00	Very strong positive
0.6–0.79	Strong positive
0.4–0.59	Moderate positive
0.2–0.39	Weak positive
0–0.19	Very weak positive
0–(−0.19)	Very weak positive
(−0.20)–(−0.39)	Weak negative
(−0.40)–(−0.59)	Moderate negative
(−0.60)–(−0.79)	Strong negative
(−0.80)–(−1.00)	Very strong negative

The $(HR)^2$ is used to interpret the proportion of the total variation in Y explained by the regression line and another way to express the level of prediction accuracy (Table 5.4).

Coefficient of determination $(HR)^2 = SSR/SST$. Therefore:

- For Y, $(HR)^2 = 1$
- For Y_L, $(HR)^2 = 0.52773$
- For Y_R, $(HR)^2 = 0.56154$.

From the result above, the model shows that water quality index (WQI) can be predicted without error from six parameters or predictors. The strength of correlation between predictors and response variable are measured using correlation coefficient. The fuzzy correlation coefficient of the model for Y_L and Y_R are $0.72645 = \sqrt{0.52773}$ and $0.74936 = \sqrt{0.56154}$, signifying a strong positive linear correlation between WQI and the predictors. Therefore, these six variables provide a reliable prediction model of WQI. Next, we propose new method that can improve the correlation coefficient of the prediction model.

5.2 An Enhanced Fuzzy Convex Combination for Improved Fuzzy Model

Since the previous fuzzy model only give about 0.72645 and 0.74935, we may improve the model by applying some convex combination to the model. The fuzzy convex combination is based on the fuzzy model to improve the result for the WQI value. The general form of the model can be written as below,

$$\tilde{Y}_i = (1 - \alpha)Y_L - \alpha Y_R$$

where Y_L is the fuzzy equation for the left spread and Y_R is the fuzzy equation for right spread with the same membership number. From our observation, we can say that:

1. When $\alpha = 0$, the model is equal to its right fuzzy equation, Y_L
2. When $\alpha = 1$, the model is equal to its right fuzzy equation, Y_R
3. When $\alpha = 0.5$, the model is equal with the crisp value i.e. standard DOE formulation.

The proposed fuzzy convex combination form for membership function, $\mu = 0$ and $\alpha = 0.4$, $\alpha = 0.6$, $\alpha = 0.45$, $\alpha = 0.55$ are given below.

$$
\tilde{Y}_{\alpha=0.4} = 0.6 \begin{bmatrix} (-1.18234\text{E}-11, 7.2863) + (0.22, -0.0318)\text{DO} \\ +(0.19, -0.0849)\text{BOD} + (0.16, 0.0549)\text{COD} \\ +(0.15, 0.0143)\text{AN} + (0.16, -0.0094)\text{SS} + (0.12, -0.0050)\text{pH} \end{bmatrix}
$$

$$
+ \; 0.4 \begin{bmatrix} (-1.18234\text{E}-11, 7.2863) + (0.22, -0.0318)\text{DO} \\ +(0.19, -0.0849)\text{BOD} + (0.16, 0.0549)\text{COD} \\ +(0.15, 0.0143)\text{AN} + (0.16, -0.0094)\text{SS} + (0.12, -0.0050)\text{pH} \end{bmatrix}
$$

$$
\tilde{Y}_{\alpha=0.6} = 0.4 \begin{bmatrix} (-1.18234\text{E}-11, 7.2863) + (0.22, -0.0318)\text{DO} \\ +(0.19, -0.0849)\text{BOD} + (0.16, 0.0549)\text{COD} \\ +(0.15, 0.0143)\text{AN} + (0.16, -0.0094)\text{SS} + (0.12, -0.0050)\text{pH} \end{bmatrix}
$$

$$
+ \; 0.6 \begin{bmatrix} (-1.18234\text{E}-11, 7.2863) + (0.22, -0.0318)\text{DO} \\ +(0.19, -0.0849)\text{BOD} + (0.16, 0.0549)\text{COD} \\ +(0.15, 0.0143)\text{AN} + (0.16, -0.0094)\text{SS} + (0.12, -0.0050)\text{pH} \end{bmatrix}
$$

$$
\tilde{Y}_{\alpha=0.45} = 0.55 \begin{bmatrix} (-1.18234\text{E}-11, 7.2863) + (0.22, -0.0318)\text{DO} \\ +(0.19, -0.0849)\text{BOD} + (0.16, 0.0549)\text{COD} \\ +(0.15, 0.0143)\text{AN} + (0.16, -0.0094)\text{SS} + (0.12, -0.0050)\text{pH} \end{bmatrix}
$$

$$
+ \; 0.45 \begin{bmatrix} (-1.18234\text{E}-11, 7.2863) + (0.22, -0.0318)\text{DO} \\ +(0.19, -0.0849)\text{BOD} + (0.16, 0.0549)\text{COD} \\ +(0.15, 0.0143)\text{AN} + (0.16, -0.0094)\text{SS} + (0.12, -0.0050)\text{pH} \end{bmatrix}
$$

$$
\tilde{Y}_{\alpha=0.55} = 0.45 \begin{bmatrix} (-1.18234\text{E}-11, 7.2863) + (0.22, -0.0318)\text{DO} \\ +(0.19, -0.0849)\text{BOD} + (0.16, 0.0549)\text{COD} \\ +(0.15, 0.0143)\text{AN} + (0.16, -0.0094)\text{SS} + (0.12, -0.0050)\text{pH} \end{bmatrix}
$$

$$
+ \; 0.55 \begin{bmatrix} (-1.18234\text{E}-11, 7.2863) + (0.22, -0.0318)\text{DO} \\ +(0.19, -0.0849)\text{BOD} + (0.16, 0.0549)\text{COD} \\ +(0.15, 0.0143)\text{AN} + (0.16, -0.0094)\text{SS} + (0.12, -0.0050)\text{pH} \end{bmatrix}
$$

From the results from the Figs. 5.1, 5.2, 5.3, 5.4 and 5.5, it shows that convex combination for $\alpha = 0.45$ and $\alpha = 0.55$ produce smaller gap from its WQI compared to convex combination value of $\alpha = 0.4$ and $\alpha = 0.6$. To strengthen the reason of choosing convex value of $\alpha = 0.55$ for this model, we analyzed three types of errors which are SSR, SSE and SST and used $(HR)^2$ to express the level of prediction accuracy.

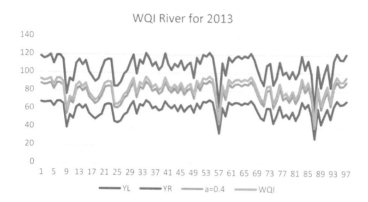

Fig. 5.1 Result for using fuzzy convex combination for $\alpha = 0.4$

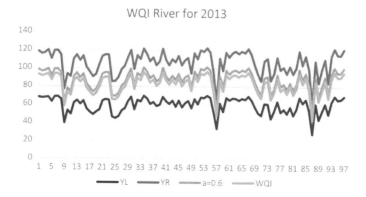

Fig. 5.2 Result for using fuzzy convex combination for $\alpha = 0.6$

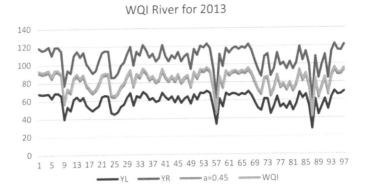

Fig. 5.3 Result for using fuzzy convex combination for $\alpha = 0.45$

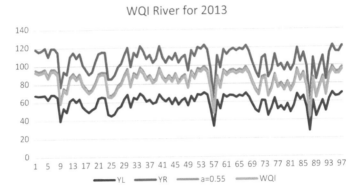

Fig. 5.4 Result for using fuzzy convex combination for $\alpha = 0.55$

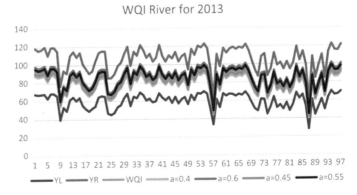

Fig. 5.5 Comparison result using convex combination of $\alpha = 0.4, \alpha = 0.6, \alpha = 0.45$ and $\alpha = 0.55$

Table 5.5 Results of error analysis

	$\alpha = 0.4$	$\alpha = 0.6$	$\alpha = 0.45$	$\alpha = 0.55$
SSR	11956.37305	13709.22125	10784.48946	11660.91356
SSE	2146.79420	2146.79420	536.69855	536.69855
SST	14103.16725	15856.01545	11321.18801	12197.6121
$(HR)^2$	0.84778	0.86461	0.95259	0.95599
HR	0.92075	0.92984	0.97601	0.97774

Based on the Table 5.5 and Fig. 5.5, we can conclude that the best convex value that can be used for prediction is $\alpha = 0.55$. The result shows that nearly 95.599% of WQI can be explained by all six predictors. As the fuzzy correlation coefficient of convex value is $0.97774 = \sqrt{0.95599}$, it clearly shown that it signifies a strong positive linear correlation between WQI and the other six predictors. Furthermore, the class of WQI obtained from the proposed model is almost the same as WQI value obtain from DOE sampling as can be seen in the following section.

5.3 Prediction of Water Quality Index

From the calculation and results from the previous section, we used $\alpha = 0.55$ as the convex value for the predictive modelling for WQI. For this model, we used the remaining rivers that haven't been used in the previous simulations. The following Table 5.6 summarizes all of the results.

The selection of the free parameter α in convex scheme is very vital in obtaining the significant values of WQI prediction. From Fig. 5.6, we can see that by using $\alpha = 0.55$, we can obtain almost the same WQI value from DOE sampling. This explain that we can calculate and predict WQI value by using fuzzy convex combination model. This shows that the proposed predictive model is able to give good prediction for WQI with higher accuracy (Table 5.7).

5.4 Conclusion

One of the main concerns in explaining the associations between predictors and response variable is opting the appropriate linear model which can reduce computation risk and giving flexible interpretations. This study enables to assess the water pollution based on water quality index (WQI).

The main purpose of employing the proposed model is on the development of fuzzy multivariate regression models to predict the value of WQI at selected rivers in the Perak, Malaysia. The main contribution of the present study is the enhancement of data preprocessing technique and the new improvement to calculate the spreading

Table 5.6 Comparison of WQI and by using convex combination

Basin	River	STA no	2013 WQI	2013 α = 0.55	2014 WQI	2014 α = 0.55	2015 WQI	2015 α = 0.55	2016 WQI	2016 α = 0.55	2017 WQI	2017 α = 0.55
Perak	Woh	2PK63	93.49235	95.98262	93.13105	95.6561	94.42776	96.974	90.10516	92.50422	91.41681	93.84574
			82.74513	84.94395	91.44174	93.86419	90.22125	92.6431	86.85158	89.0948	84.37395	86.69615
			90.66687	93.08184	92.96758	95.45862	95.18061	97.75579	92.73361	95.23106	92.90855	95.43923
					94.02856	96.57412	88.78752	91.26494	95.35974	97.95956	90.1275	92.53146
											90.64212	93.03544
	Seluang	2PK52	75.08277	77.15326	38.76594	39.97866	56.09562	57.74429	59.73113	61.63426	57.74696	59.73222
			65.03248	66.948	58.95115	60.69647	62.49344	64.07192	63.92126	65.7536	69.07693	71.24197
			37.91183	39.04393	50.45383	51.87533	54.7401	56.59008	73.28881	75.33345	52.5814	54.3929
			66.25996	68.19397	52.37064	53.91294	71.72278	73.64445			48.6237	50.3536
	Sungkai	2PK07	91.07837	93.61411	86.83359	89.16574	91.90424	94.33581	86.98535	89.33444	84.80035	87.17915
			90.85026	93.37276	90.21549	92.62629	94.6293	97.1761	92.43429	94.95162	88.61959	91.07235
			86.40236	88.63891	81.40736	83.45993	92.00348	94.47408	87.93242	90.26615	89.15837	91.54969
			92.18875	94.65411	92.44305	94.92387	87.32167	89.70611	82.96273	85.36707	83.93216	86.13659

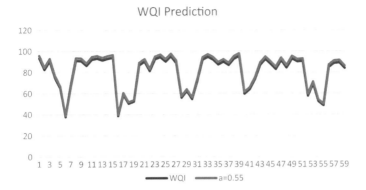

Fig. 5.6 Prediction result by using convex value of $\alpha = 0.55$

Table 5.7 WQI list of class prediction for fuzzy convex combination	WQI		a = 0.55	
	93.49235	I	95.98262	I
	82.74513	II	84.94395	II
	90.66687	II	93.08184	I
	75.08277	III	77.15326	II
	65.03248	III	66.948	III
	37.91183	IV	39.04393	IV
	66.25996	III	68.19397	III
	91.07837	II	93.61411	I
	90.85026	II	93.37276	I
	86.40236	II	88.63891	II
	92.18875	II	94.65411	I
	93.13105	I	95.6561	I
	91.44174	II	93.86419	I
	92.96758	I	95.45862	I
	94.02856	I	96.57412	I
	38.76594	IV	39.97866	IV
	58.95115	III	60.69647	III
	50.45383	IV	51.87533	IV
	52.37064	III	53.91294	III
	86.83359	II	89.16574	II
	90.21549	II	92.62629	II
	81.40736	II	83.45993	II
	92.44305	II	94.92387	I

(continued)

Table 5.7 (continued)

WQI		a = 0.55	
94.42776	I	96.974	I
90.22125	II	92.6431	II
95.18061	I	97.75579	I
88.78752	II	91.26494	II
56.09562	III	57.74429	III
62.49344	III	64.07192	III
54.7401	III	56.59008	III
71.72278	III	73.64445	III
91.90424	II	94.33581	I
94.6293	I	97.1761	I
92.00348	II	94.47408	I
87.32167	II	89.70611	II
90.10516	II	92.50422	II
86.85158	II	89.0948	II
92.73361	I	95.23106	I
95.35974	I	97.95956	I
59.73113	III	61.63426	III
63.92126	III	65.7536	III
73.28881	III	75.33345	III
86.98535	II	89.33444	II
92.43429	II	94.95162	I
87.93242	II	90.26615	II
82.96273	II	85.36707	II
91.41681	II	93.84574	I
84.37395	II	86.69615	II
92.90855	I	95.43923	I
90.1275	II	92.53146	II
90.64212	II	93.03544	I
57.74696	III	59.73222	III
69.07693	III	71.24197	III
52.5814	III	54.3929	III
48.6237	IV	50.3536	IV
84.80035	II	87.17915	II
88.61959	II	91.07235	II
89.15837	II	91.54969	II
83.93216	II	86.13659	II

value. For data preprocessing, fuzzy convex combination has been proposed. The proposed methods have proven to be able to increase water quality result (WQI) accuracy compared to existing methods. Other than that, the predicting models produced in this research includes fuzzy multivariate regression and convex combination method have been successfully tested and validated with the real data. Hence, the proposed fuzzy multivariate regression model and convex combination model is the best to use to calculate WQI and predict WQI value compared to some existing methods.

References

1. Abdullah L, Zakaria N (2012) Matrix driven multivariate fuzzy linear regression model in car sales. J Appl Sci (Faisalabad) 12(1):56–63
2. Chang PT, Lee ES (1996) A generalized fuzzy weighted least-squares regression. Fuzzy Sets Syst 82(3):289–298
3. Chowdhury AK, Debsarkar A, Chakrabarty S (2015) Novel methods for assessing urban air quality: combined air and noise pollution approach. J Atmos Pollut 3(1):1–8
4. Hidayah Mohamed Isa N, Othman M, Karim SAA (2018) Multivariate matrix for fuzzy linear regression model to analyse the taxation in Malaysia. Int J Eng Technol 7(4.33):78–82
5. Pan NF, Lin TC, Pan NH (2009) Estimating bridge performance based on a matrix-driven fuzzy linear regression model. Autom Const 18(5):578–586

Index

© The Author(s), under exclusive licence to Springer Nature Singapore Pte Ltd. 2020
S. A. A. Karim and N. F. Kamsani, *Water Quality Index Prediction Using Multiple Linear Fuzzy Regression Model*, SpringerBriefs in Water Science and Technology,
https://doi.org/10.1007/978-981-15-3485-0